Estimates of Tracer-Based Piston-Flow Ages of Groundwater from Selected Sites: National Water-Quality Assessment Program, 2006–10

By Stephanie D. Shapiro, L. Niel Plummer, Eurybiades Busenberg, Peggy K. Widman, Gerolamo C. Casile, Julian E. Wayland, and Donna L. Runkle

National Water-Quality Assessment Program

Scientific Investigations Report 2012–5141

U.S. Department of the Interior
U.S. Geological Survey

U.S. Department of the Interior
KEN SALAZAR, Secretary

U.S. Geological Survey
Marcia K. McNutt, Director

U.S. Geological Survey, Reston, Virginia: 2012

For more information on the USGS—the Federal source for science about the Earth, its natural and living resources, natural hazards, and the environment, visit http://www.usgs.gov or call 1–888–ASK–USGS.

For an overview of USGS information products, including maps, imagery, and publications, visit http://www.usgs.gov/pubprod

To order this and other USGS information products, visit http://store.usgs.gov

Suggested citation:
Shapiro, S.D., Plummer, L.N., Busenberg, E., Widman, P.K., Casile, G.C., Wayland, J.E., and Runkle, D.L., 2012, Estimates of tracer-based piston-flow ages of groundwater from selected sites—National Water-Quality Assessment Program, 2006–10: U.S. Geological Survey Scientific Investigations Report 2012–5141, 100 p.

Foreword

The U.S. Geological Survey (USGS) is committed to providing the Nation with reliable scientific information that helps to enhance and protect the overall quality of life and that facilitates effective management of water, biological, energy, and mineral resources (http://www.usgs.gov/). Information on the Nation's water resources is critical to ensuring long-term availability of water that is safe for drinking and recreation and is suitable for industry, irrigation, and fish and wildlife. Population growth and increasing demands for water make the availability of that water, measured in terms of quantity and quality, even more essential to the long-term sustainability of our communities and ecosystems.

The USGS implemented the National Water-Quality Assessment (NAWQA) Program in 1991 to support national, regional, State, and local information needs and decisions related to water-quality manage¬ment and policy (http://water.usgs.gov/nawqa). The NAWQA Program is designed to answer: What is the quality of our Nation's streams and groundwater? How are conditions changing over time? How do natural features and human activities affect the quality of streams and groundwater, and where are those effects most pronounced? By combining information on water chemistry, physical characteristics, stream habitat, and aquatic life, the NAWQA Program aims to provide science-based insights for current and emerging water issues and priorities. From 1991 to 2001, the NAWQA Program completed interdisciplinary assessments and established a baseline understanding of water-quality conditions in 51 of the Nation's river basins and aquifers, referred to as Study Units (http://water.usgs.gov/nawqa/studies/study_units.html).

National and regional assessments are ongoing in the second decade (2001–2012) of the NAWQA Program as 42 of the 51 Study Units are selectively reassessed. These assessments extend the findings in the Study Units by determining water-quality status and trends at sites that have been consistently monitored for more than a decade, and filling critical gaps in characterizing the quality of surface water and groundwater. For example, increased emphasis has been placed on assessing the quality of source water and finished water associated with many of the Nation's largest community water systems. During the second decade, NAWQA is addressing five national priority topics that build an understanding of how natural features and human activities affect water quality, and establish links between sources of contaminants, the transport of those contaminants through the hydrologic system, and the potential effects of contaminants on humans and aquatic ecosystems. Included are studies on the fate of agricultural chemicals, effects of urbanization on stream ecosystems, bioaccumulation of mercury in stream ecosystems, effects of nutrient enrichment on aquatic ecosystems, and transport of contaminants to public-supply wells. In addition, national syntheses of information on pesticides, volatile organic compounds (VOCs), nutrients, trace elements, and aquatic ecology are continuing.

The USGS aims to disseminate credible, timely, and relevant science information to address practical and effective water-resource management and strategies that protect and restore water quality. We hope this NAWQA publication will provide you with insights and information to meet your needs, and will foster increased citizen awareness and involvement in the protection and restoration of our Nation's waters.

The USGS recognizes that a national assessment by a single program cannot address all water-resource issues of interest. External coordination at all levels is critical for cost-effective management, regulation, and conservation of our Nation's water resources. The NAWQA Program, therefore, depends on advice and information from other agencies—Federal, State, regional, interstate, Tribal, and local—as well as nongovernmental organizations, industry, academia, and other stakeholder groups. Your assistance and suggestions are greatly appreciated.

William H. Werkheiser
USGS Associate Director for Water

Contents

Figures

Table

Conversion Factors, Datums, and Abbreviations and Acronyms

Conversion Factors

Inch/Pound to SI

Multiply	By	To obtain
	Length	
foot (ft)	0.3048	meter (m)
mile (mi)	1.609	kilometer (km)

SI to Inch/Pound

Multiply	By	To obtain
	Volume	
liter (L)	0.2642	gallon (gal)
cubic centimeter (cm^3)	0.06102	cubic inch (in^3)
	Mass	
gram (g)	0.03527	ounce, avoirdupois (oz)
kilogram (kg)	2.205	pound avoirdupois (lb)

Temperature in degrees Celsius (°C) may be converted to degrees Fahrenheit (°F) as follows:

$$°F=(1.8×°C)+32.$$

Specific conductance is given in microsiemens per centimeter at 25 degrees Celsius (µS/cm at 25°C).

Concentrations of most chemical constituents in water are given either in milligrams per liter (mg/L) or micrograms per liter (µg/L).

Chlorofluorocarbon (CFC) concentrations are given in units of picograms per kilogram (pg/kg) and picomoles per kilogram (pmol/kg). One picogram is 10^{-12} grams. One picomole is 10^{-12} moles. One mole contains 6.022×10^{23} atoms or molecules of a substance. Sulfur hexafluoride (SF_6) concentrations are given in units of femtograms per kilogram (fg/kg) and femtomoles per kilogram (fmol/kg). One femtogram is 10^{-15} grams. One femtomole is 10^{-15} moles. CFC and SF_6 concentrations also are given in units of parts per trillion by volume (pptv), which represents the atmospheric concentration that would have yielded a measured aqueous concentration assuming equilibrium partitioning between atmosphere and water under the specified conditions (recharge temperature, recharge elevation, and excess air concentration).

Tritium concentrations are given in units of Tritium Units (TU). Based upon a tritium half-life of 12.32 years (Lucas and Unterweger, 2000), 1 TU is equal to 3.22 picocuries per liter.

Helium-3 (^{3}He) data are reported as δ values computed from the formula

$$\delta^3 He = \left[\left(\frac{R_x}{R_{STD}}\right)-1\right]100$$

where R_x is the ratio of ^{3}He to ^{4}He in the sample, RSTD is the ^{3}He to ^{4}He ratio of the reference standard air (1.384×10^{-6}), and δ^{3}He is expressed in percent.

Conversion Factors, Datums, and Abbreviations and Acronyms—Continued

Excess air concentrations (see Glossary) are reported in units of cubic centimeters at standard temperature and pressure (STP) per kilogram of water (cc STP/kg), and helium (He) and neon (Ne) concentrations are reported in units of cubic centimeters (at standard temperature and pressure) per gram of water (cc STP/g). One cc STP of He or Ne is equal to 2.6868×10^{19} atoms.

Datum

Vertical coordinate information is referenced to the National Geodetic Vertical Datum of 1929 (NGVD 29).

Altitude, as used in this report, refers to distance above the vertical datum.

Abbreviations and Acronyms

Ar	argon
^{14}C	carbon-14
CFC	chlorofluorocarbon
CH_4	methane
CO_2	carbon dioxide
DWH	NAWQA Data Warehouse
Fe	Iron
FSS	Flow System Study
FY	fiscal year
He	helium
^3H	tritium
^3He	helium-3
^4He	helium-4
LUS	Land-Use Study
MAAT	mean annual air temperature
MAS	Major-Aquifer Study
Mn	Manganese
N_2	nitrogen (in the form dnitrogen)
NAWQA	National Water-Quality Assessment Program
Ne	neon
O_2	oxygen (in the form dioxygen)
REF	Reference well
SF_6	sulfur hexafluoride
SIO	Scripps Institution of Oceanography
SIP	standard temperature and pressure
SUS	Study-Unit SUrvey
TU	tritium unit
UA	unfractionated air
USGS	U.S. Geological Survey

Estimates of Tracer-Based Piston-Flow Ages of Groundwater from Selected Sites: National Water-Quality Assessment Program, 2006–10

By Stephanie D. Shapiro, L. Niel Plummer, Eurybiades Busenberg, Peggy K. Widman, Gerolamo C. Casile, Julian E. Wayland, and Donna L. Runkle

Abstract

Piston-flow age dates were interpreted from measured concentrations of environmental tracers from 812 National Water-Quality Assessment (NAWQA) Program groundwater sites from 27 Study Units across the United States. The tracers of interest include chlorofluorocarbons (CFCs), sulfur hexafluoride (SF_6), and tritium/helium-3 ($^3H/^3He$).

Tracer data compiled for this analysis were collected from 2006 to 2010 from groundwater wells in NAWQA studies, including:

- Land-Use Studies (LUS, shallow wells, usually monitoring wells, located in recharge areas under dominant land-use settings),

- Major-Aquifer Studies (MAS, wells, usually domestic supply wells, located in principal aquifers and representing the shallow drinking water supply),

- Flow System Studies (FSS, networks of clustered wells located along a flowpath extending from a recharge zone to a discharge zone, preferably a shallow stream) associated with Land-Use Studies, and

- Reference wells (wells representing groundwater minimally impacted by anthropogenic activities) also associated with Land-Use Studies.

Tracer data were evaluated using documented methods and are presented as aqueous concentrations, equivalent atmospheric concentrations (for CFCs and SF_6), and tracer-based piston-flow ages. Selected ancillary data, such as redox data, well-construction data, and major dissolved-gas (N_2, O_2, Ar, CH_4, and CO_2) data, also are presented. Recharge temperature was inferred using climate data (approximated by mean annual air temperature plus 1°C [MAAT +1°C]) as well as major dissolved-gas data (N_2-Ar-based) where available. The N_2-Ar-based temperatures showed significantly more variation than the climate-based data, as well as the effects of denitrification and degassing resulting from reducing conditions. The N_2-Ar-based temperatures were colder than the climate-based temperatures in networks where recharge was limited to the winter months when evapotranspiration was reduced.

The tracer-based piston-flow ages compiled in this report are provided as a consistent means of reporting the tracer data. The tracer-based piston-flow ages may provide an initial interpretation of age in cases in which mixing is minimal and may aid in developing a basic conceptualization of groundwater age in an aquifer. These interpretations are based on the assumption that tracer transport is by advection only and that no mixing occurs. In addition, it is assumed that other uncertainties are minimized, including tracer degradation, sorption, contamination, or fractionation, and that terrigenic (natural) sources of tracers, and spatially variable atmospheric tracer concentrations are constrained.

Introduction

The age of groundwater is defined as the time required for a water molecule to travel from a point of recharge to a measurement point such as a well. An understanding of groundwater age can be used to infer groundwater flowpaths and rates of recharge or biogeochemical reactions, reconstruct contaminant loading histories, explain trends in groundwater quality, gain insight into groundwater susceptibility to contamination, and constrain groundwater flow and transport models. Environmental tracers are ideal tools to estimate ages because they have a global signature, and their input histories to aquifers typically can be constrained. In this report, reference to environmental tracers pertains specifically to the use of chlorofluorocarbons (CFCs) (Busenberg and Plummer, 1992; Plummer and Busenberg, 2000; International Atomic Energy Agency, 2006; Hinkle and others, 2010), sulfur hexafluoride (SF_6) (Busenberg and Plummer, 2000; International Atomic Energy Agency, 2006), and the combination of tritium/helium-3 ($^3H/^3He$) (Schlosser and others, 1989; Solomon and Cook, 2000). Tracer ages are interpreted using the assumption of piston-flow conditions and can provide an initial interpretation of age structure in an aquifer, but also provide a practical means of reporting tracer concentration data. The use of tracer-based piston flow ages is discussed in detail in Hinkle and others (2010). The approach used here is identical to that of Hinkle and others (2010).

The National Water-Quality Assessment (NAWQA) Program of the U.S. Geological Survey (USGS) is tasked with: (1) describing the status of and trends in water quality of large, representative portions of the Nation's water resources, and (2) providing an understanding of natural and anthropogenic factors affecting the quality of these resources (Gilliom and others, 1995). Tracer-based piston-flow ages are an important component in the effort to achieve these goals.

The NAWQA Program is composed of geographically and hydrologically distinct Study Units located throughout the United States. Study Units and bibliographies of Study-Unit publications are described in detail in U.S. Geological Survey (2010).

Groundwater assessments within Study Units include networks (groups related by commonality of targeted resource) of 20–30 randomly distributed wells. These networks include wells from Major-Aquifer Studies (MASs), Land-Use Studies (LUSs), Flowpath Studies (FPSs), and also include Reference (REF) wells. These networks are described in detail in Hinkle and others (2010).

Purpose and Scope

Tracer data have been collected from NAWQA groundwater sites since 1992. The NAWQA Program has a need for: (1) a comprehensive listing of known tracer datasets, and (2) a compilation of estimates of groundwater age for these groundwater samples. One such compilation was published by Hinkle and others (2010), and includes NAWQA data from fiscal years 1992–2005. [The Federal fiscal year (FY) begins October 1 and ends September 30 of the year with which it is numbered.] This report summarizes tracer data that were collected from NAWQA LUS, MAS, FPS, and REF wells from FY 2006–10 during the second decadal cycle of the NAWQA Program. A total of 812 sites are included in this report: 437 LUS, 281 MAS, 72 FPS, and 21 REF sites, plus 1 dual-purpose LUS/FPS site.

Only samples from wells were assembled for this report; samples from springs were not included because of an almost exclusive programmatic focus on wells for groundwater sampling (Gilliom and others, 1995).

The tracers of interest include CFCs (CFC-11, CFC-12, and CFC-113), SF_6, and $^3H/^3He$, which are useful in determining tracer-based piston-flow ages on the order of years to decades. In addition, data on the major dissolved gases (N_2, O_2, Ar, CH_4, and CO_2) were assembled where available because they are used to constrain recharge temperatures and excess air concentrations.

The tracer data compiled in this report represent the CFC, SF_6, and $^3H/^3He$ data associated with the targeted (LUS, MAS, FPS, and REF) wells and collected during the targeted timeframe (FY 2006–10) from NAWQA Study Units. These Study Units are shown in figure 1 and are listed in table A1 (appendix A).

In the following sections, the tracer datasets that were assembled for this report are described, and the tracer-based piston-flow ages are provided. In addition, some insights on recharge temperatures that resulted from this compilation are briefly discussed.

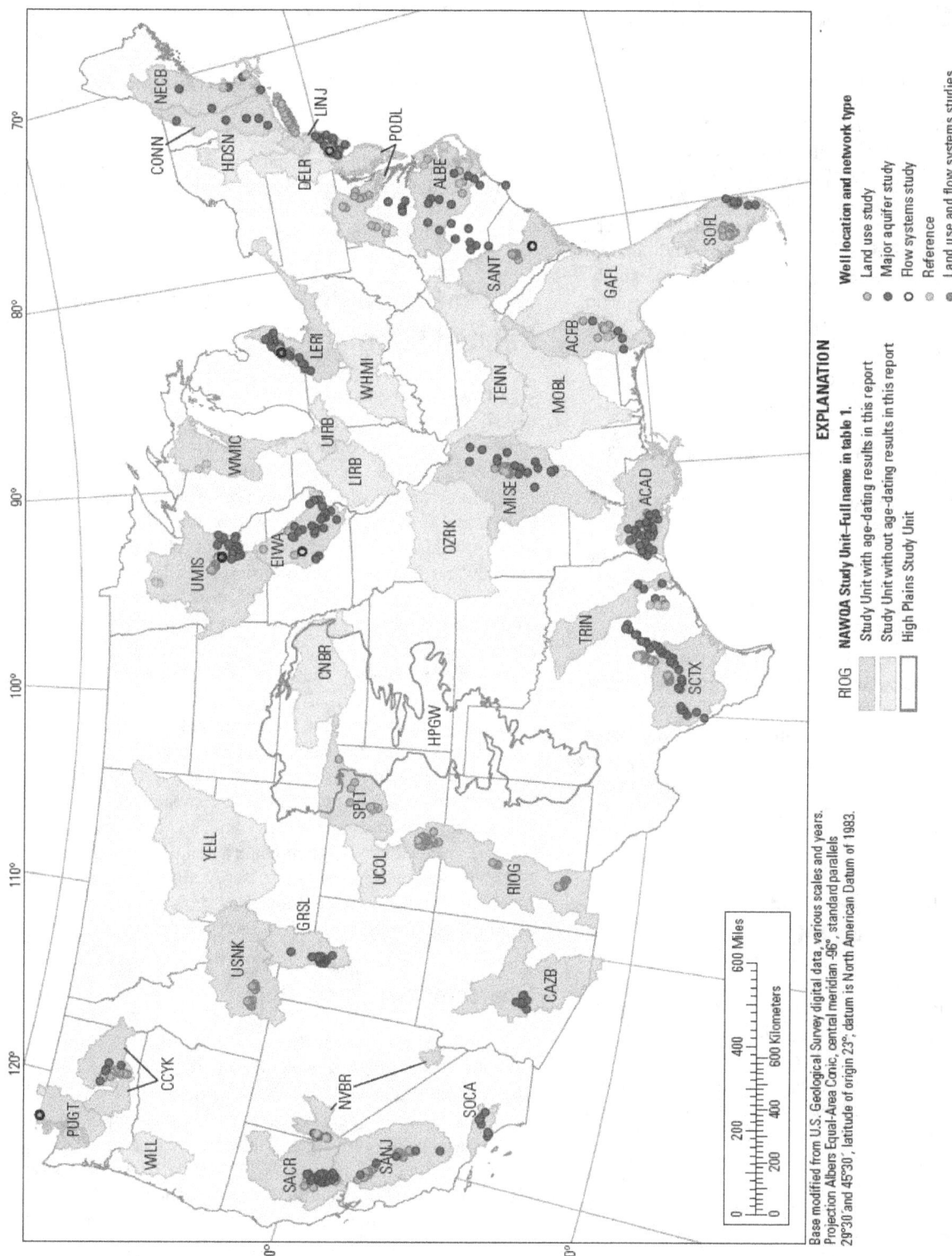

Base modified from U.S. Geological Survey digital data, various scales and years.
Projection Albers Equal-Area Conic, central meridian -96°, standard parallels
29°30' and 45°30', latitude of origin 23°; datum is North American Datum of 1983.

Figure 1. Locations of National Water-Quality Assessment Program Study Units. (Study-Unit abbreviations are defined in table A1.)

Approach Used in this Compilation

Environmental Tracer Data

Tracer data collected from FY 2006 to FY 2010 from 812 Land-Use Study (LUS), Major-Aquifer Study (MAS), Flowpath Study (FPS), and Reference (REF) sites were identified and compiled (table A1, appendix A). Analytical methods are described elsewhere (Busenberg and Plummer, 1992; Plummer and Mullin, 1997; Busenberg and Plummer, 2000). For internal consistency, CFC concentrations, which were originally measured over a range of analysis dates using a variety of calibration standards, were rescaled to the Scripps Institution of Oceanography 2005 scale of atmospheric CFC mixing ratios (Walker and others, 2009). All SF_6 data were scaled to the National Oceanic and Atmospheric Administration 2000 scale of atmospheric SF_6 mixing ratios (National Oceanic and Atmospheric Administration, 2008).

For tracer data interpretation, the processes used are documented in detail in Hinkle and others (2010) and are not repeated here. The same procedure was used in this compilation in order to make the two reports consistent. One minor difference was that SF_6 concentrations from repeat samples from the same well were averaged to determine an age, instead of averaging the apparent ages of the two samples. This approach was used to be consistent with the way that the apparent CFC ages were determined for multiple samples from the same well. The tracer-based ages may approximate time-of-travel in certain physical settings; however, a more detailed knowledge of the hydrogeologic environment, as well as the development of mixing models, typically are needed to provide more realistic age interpretations. This kind of effort is beyond the scope of this report.

Ancillary Chemistry, Water Level, Well Construction, and Tritium Data

Ancillary chemistry, water level, well construction, and ^3H data for individual sites were obtained from the NAWQA Data Warehouse (DWH) (U.S. Geological Survey, 2011).

Water level and ancillary chemistry data typically were from the same sampling date as the environmental tracers, but in those cases where the date was not the same, a notation was made in the tables in appendix B. The ancillary chemistry data provided insight into redox conditions in the aquifer, particularly if major dissolved-gas data were not available.

Estimates of spatial and temporal variations in ^3H in precipitation were used for reconstructed ^3H analysis. Reconstructed ^3H analysis is based on back-decaying measured ^3H concentrations in groundwater, using tracer-based piston-flow ages to estimate the time-dependent amount of ^3H decay, and comparing these undecayed or original ^3H concentrations to historical ^3H inputs (see for example: Dunkle and others, 1993; Ekwurzel and others,

1994). Estimates of ^3H in precipitation were based on International Atomic Energy Agency data from 10 long-term stations in the conterminous United States. Missing temporal records were replaced with values based on correlation using long-term datasets from Ottawa, Canada (August 1953–December 1987) and Vienna, Austria (January 1988–December 2001). Concentrations prior to August 1953 were estimated using graphical pre-bomb distributions of Thatcher (1962).

Tracer-Based Piston-Flow Ages and Related Information

This section documents newly interpreted tracer data for 812 sites evaluated as part of this compilation. In addition, some insights about recharge temperatures, derived from analysis of major dissolved-gas data, are briefly discussed.

Interpretations of tracer data are given as tracer-based piston-flow ages and as tracer-based piston-flow recharge dates. As addressed in Hinkle and others (2010), a groundwater sample collected during calendar year 2000 and attributed with a tracer-based piston-flow age of 10 years would have a tracer-based piston-flow recharge date of (calendar year) 1990. Tracer-based piston-flow ages and tracer-based piston-flow recharge dates often are censored with a ">" (greater than) or "<" (less than). For example, if CFC degradation appeared to be present in a sample, the tracer-based piston-flow age and tracer-based piston-flow recharge date could have an old bias. The tracer-based piston-flow recharge date could be accompanied by a ">" to indicate a greater (more recent) date, and the tracer-based piston-flow age by a "<" to indicate a smaller (younger) age. Sites with age-dating results that have been censored with ">" or "<" do not have discrete tracer-based piston-flow ages or discrete tracer-based piston-flow recharge dates.

Environmental Tracer Data

Environmental tracer data from 812 sites in 27 Study Units were interpreted here as tracer-based piston-flow ages. The data are organized by Study Unit and network.

The measured tracer concentration data, interpreted tracer-based piston-flow ages, and ancillary data for each network are reported in appendix B. The derived tracer-based piston-flow ages are examined for consistency with local age gradients and tritium data, and also are compared against each other when multiple tracers are available for a given site. A summary of the tracer-based piston-flow ages by sample is given in table 1 (at back of report). Where tracer-based piston-flow ages from more than one tracer type (CFCs, SF_6, $^3H/^3He$) were available for a given site, interpretations from all tracers are provided.

Major-Dissolved-Gas-Based and Climate-Based Recharge Temperatures

A compilation of major dissolved-gas data that were used for estimation of recharge temperatures is included wherever such samples were taken. Recharge temperature often is assumed to be close to either the MAAT (Andrews, 1992) or the MAAT +1°C (Stute and Schlosser, 2000) (MAAT +1°C was used in this report). In many networks, the recharge temperatures based on major dissolved-gas data are comparable to MAAT +1°C (fig. 2). Considering sites from aquifers composed of sediments, the differences between the recharge temperatures based on major dissolved-gas data and those based on MAAT +1°C were, on average, about 0.54°C (n=449). However, the standard deviation of these differences was 4.3°C. As noted by Hinkle and others (2010), climate data are useful for estimation of average recharge temperatures; however, recharge temperatures vary greatly around this average, and characterization of site-specific recharge temperatures benefits from site-specific data, such as N_2-Ar-data, as long as these data are not affected by highly reduced conditions and gas-stripping. The uncertainty in the

estimation of recharge temperature can lead to uncertainty in the tracer-based piston-flow ages that varies in magnitude depending on the age of the water sample (Plummer and Busenberg, 2000).

In this compilation, the climate data are significantly warmer for seven networks (ACAD, ALBE, CONN, LINJ, NECB, POTO, and SANA) than the N_2-Ar-inferred recharge temperatures and are shifted to the left of the 1:1 line in figure 2. For these seven networks, samples are taken from wells in locations where recharge likely occurs in the winter months when evapotranspiration is minimized. For these locations, the N_2-Ar-inferred recharge temperatures would be colder than the MAAT+1°C. In locations that are not dominated by seasonal growth, N_2-Ar-inferred recharge temperatures and climate-based temperatures are more comparable. The large spread in N_2-Ar-inferred recharge temperatures for a network, such as UMIS, likely results from the mixture of oxic and suboxic conditions [identified where water contained O_2 < 1 mg/L, Mn > 50 µg/L, Fe >100 µg/L, and(or) CH_4 > 1 µg/L as discussed in Hinkle and others (2010)], as well as probable gas stripping issues.

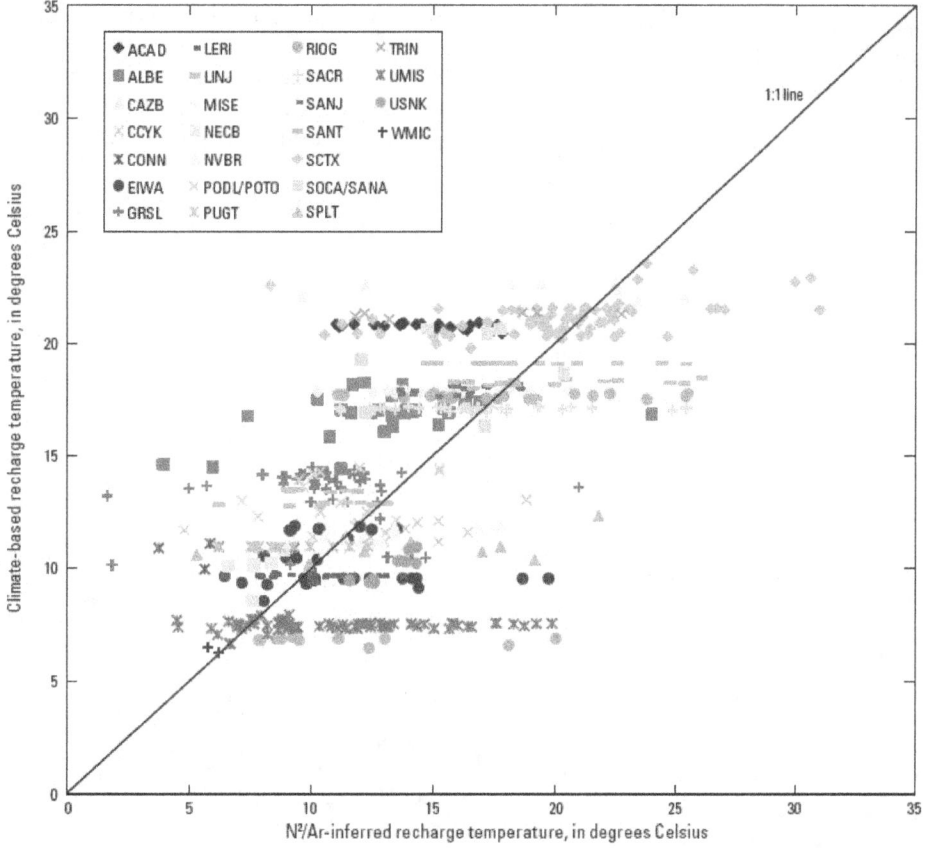

Figure 2.   Comparison of N_2-Ar-inferred recharge temperatures and climate-based recharge temperatures (mean annual air temperature + 1°C). Study-Unit abbreviations are defined in table A1. N_2-Ar-inferred recharge temperatures and climate-based recharge temperatures are listed for individual sites in appendix B.

Summary and Conclusions

CFC, SF_6, and $^3H/^3He$ tracer data collected from 812 NAWQA Program groundwater sites across the United States were compiled in this report. The data were from Land-Use Study (LUS), Major-Aquifer Study (MAS), Flowpath Studies (FPS) and Reference (REF) networks. The time period focused on Federal fiscal years 2006–10. Tracer data from other NAWQA Program components were not compiled here.

Tracer data were evaluated using established methods and are presented as aqueous concentrations, equivalent atmospheric concentrations (for CFCs and SF_6), and tracer-based piston-flow ages, and also include selected ancillary data, such as redox data, well-construction data, and major dissolved-gas (N_2, O_2, Ar, CH_4, and CO_2) data. Brief summaries of each tracer dataset also are included.

Recharge temperature often is inferred using climate data (approximated by MAAT +1°C); however, these temperatures typically only represent an average recharge temperature, around which there can be significant variation as determined using the N_2-Ar-based data. For aquifers composed of sediments in this compilation, differences between N_2-Ar-based recharge temperatures and recharge temperatures based on MAAT+1°C were, on average, within about 0.54°C; however, the standard deviation of these differences was 4.3°C. In this compilation, not only did N_2-Ar-based temperatures show significantly more variation than climate-based temperatures, but they also showed entire networks where recharge is limited to the winter months when evapotranspiration is limited and the climate data are significantly warmer than the N_2-Ar-inferred recharge temperatures. The N_2-Ar-based temperatures also showed effects of degassing resulting from highly reduced conditions in numerous networks. The degassing, usually by excess N_2, can partially strip CFCs, SF_6, and ^3He resulting in an old bias in CFC and SF_6 ages and young bias in $^3H/^3He$ ages.

An understanding of groundwater age can be used to (1) infer groundwater flowpaths and rates of recharge or biogeochemical reactions, (2) reconstruct contaminant loading histories, (3) explain trends in groundwater quality, (4) gain insight into groundwater susceptibility to contamination, and (5) constrain groundwater flow and transport models. Characterizing groundwater age using tracers, however, is a complex undertaking. As evidenced in this compilation, as well as that of Hinkle and others (2010), collection of tracer data does not necessarily lead to determination of groundwater age. Collecting tracer data is only part of the investigation, but can be of greater use if multiple tracers are utilized. In addition, the outcome can be improved if the selected tracers are relevant to the area that is sampled (that is, are not degraded, do not have additional unconstrained sources, etc.), additional tracers are included that would provide age estimates on the timescale of centuries or millennia for wells with mixtures of very old and very young water, ancillary geochemical data are collected, and the study design is carefully considered (attempting to avoid wells with large open holes, heavily pumped wells, etc.).

References Cited

Andrews, J.N., 1992, Mechanisms for noble gas dissolution by groundwaters, *in* International Atomic Energy Agency, ed., Isotopes of Noble Gases as Tracers in Environmental Studies: Vienna, International Atomic Energy Agency, p. 87–110.

Busenberg, Eurybiades, and Plummer, L.N., 1992, Use of chlorofluorocarbons (CCl_3F and CCl_2F_2) as hydrologic tracers and age-dating tools—The alluvium and terrace system of central Oklahoma: Water Resources Research, v. 28, p. 2,257–2,283.

Busenberg, Eurybiades, and Plummer, L.N., 2000, Dating young groundwater with sulfur hexafluoride—Natural and anthropogenic sources of sulfur hexafluoride: Water Resources Research, v. 36, p. 3,011–3,030.

Dunkle, S.A., Plummer, L.N., Busenberg, Eurybiades, Phillips, P.J., Denver, J.M., Hamilton, P.A., Michel, R.L., and Coplen, T.B., 1993, Chlorofluorocarbons (CCl_3F and CCl_2F_2) as dating tools and hydrologic tracers in shallow groundwater of the Delmarva Peninsula, Atlantic Coastal Plain, United States: Water Resources Research, v. 29, p. 3,837–3,860.

Ekwurzel, Brenda, Schlosser, Peter, Smethie, W.M., Jr., Plummer, L.N, Busenberg, Eurybiades, Michel, R.L., Weppernig, Ralf, and Stute, Martin, 1994, Dating of shallow groundwater—Comparison of the transient tracers $^3H/^3He$, chlorofluorocarbons, and ^{85}Kr: Water Resources Research, v. 30, p. 1,693–1,708.

Gilliom, R.J., Alley, W.M., and Gurtz, M.E., 1995, Design of the National Water-Quality Assessment Program—Occurrence and distribution of water-quality conditions: U.S. Geological Survey Circular 1112, 33 p.

Hinkle, S.R., Shapiro, S.D., Plummer, L.N., Busenberg, Eurybiades, Widman, P.K., Casile, G.C., and Wayland, J.E., 2010, Estimates of tracer-based piston-flow ages of groundwater from selected sites—National Water-Quality Assessment Program, 1992–2005: U.S. Geological Survey Scientific Investigations Report 2010–5229, 90 p. (Also available at http://pubs.usgs.gov/sir/2010/5229/.)

International Atomic Energy Agency, 2006, Use of chlorofluorocarbons in hydrology—A Guidebook: Vienna, International Atomic Energy Agency, 277 p.

National Oceanic and Atmospheric Administration, 2008, NOAA calibration scales for various trace gases: National Oceanic and Atmospheric Administration, accessed December 12, 2011, at http://www.esrl.noaa.gov/gmd/ccl/scales.html.

Plummer, L.N., and Busenberg, Eurybiades, 2000, Chlorofluorocarbons, *in* Cook, P.G., and Herczeg, A.L., eds., Environmental Tracers in Subsurface Hydrology: Boston, Kluwer, p. 441–478.

Plummer, L.N., and Mullin, A.H., 1997, Collection, processing, and analysis of ground-water samples for tritium/helium-3 dating: U.S. Geological Survey National Water Quality Laboratory Technical Memorandum 97.04S, accessed December 7, 2011, at http://nwql.usgs.gov/Public/tech_memos/nwql.1997-04S.pdf.

Schlosser, Peter, Stute, Martin, Sonntag, Christian, and Münnich, K.O., 1989, Tritiogenic ^3He in shallow groundwater: Earth and Planetary Science Letters, v. 94, p. 245–256.

Solomon, D.K., and Cook, P.G., 2000, ^3H and ^3He, *in* Cook, P.G., and Herczeg, A.L., eds., Environmental Tracers in Subsurface Hydrology: Boston, Kluwer, p. 397–424.

Stute, Martin, and Schlosser, Peter, 2000, Atmospheric noble gases, *in* Cook, P.G., and Herczeg, A.L., eds., Environmental Tracers in Subsurface Hydrology: Boston, Kluwer, p. 349–377.

Thatcher, L.L., 1962, The distribution of tritium fallout in precipitation over North America: Bulletin of the International Association of Scientific Hydrology, v. 7, p. 48–58.

U.S. Geological Survey, 2010, Complete listing of NAWQA's study units and summary reports: website, accessed December 12, 2011, at http://water.usgs.gov/nawqa/studies/study_units_listing.html.

U.S. Geological Survey, 2011, USGS National Water Quality Assessment Data Warehouse: website, accessed December 12, 2011, at http://infotrek.er.usgs.gov/apex/f?p=NAWQA:HOME:0.

Walker, S.J., Weiss, R.F., and Salameh, P.K., 2009, Reconstructed histories of the annual mean atmospheric mole fractions for the halocarbons CFC-11, CFC-12, CFC-113 and carbon tetrachloride: University of California San Diego, accessed December 12, 2011, at http://bluemoon.ucsd.edu/pub/cfchist/.

Glossary

^{3}H/^{3}He ^{3}H/^{3}He refers to the combined use of ^{3}H (tritium) and its decay product, ^{3}He (helium-3). Radioactive decay of ^{3}H to ^{3}He allows elapsed time to be calculated by comparing the tritiogenic ^{3}He (that is, the amount of ^{3}He that is attributed to decay of ^{3}H) to the original amount of ^{3}H, where the original amount of ^{3}H is equal to the measured ^{3}H plus the tritiogenic ^{3}He.

^{4}He   The amount of ^{4}He (helium-4) in a water sample that is in excess to that attributed to solubility equilibrium with air, expressed as a percentage of the ^{4}He in water at solubility equilibrium. Δ^4He usually is from excess air and from elemental radioactive decay such as decay of uranium and thorium in rocks.

Ne   The amount of Ne (neon) in a water sample that is in excess to that attributed to solubility equilibrium with air, expressed as a percentage of the Ne in water at solubility equilibrium. Ne usually is from excess air.

Anthropogenic   Resulting from or pertaining to human activities.

CFCs   CFCs, or chlorofluorocarbons, are anthropogenic compounds that have been commercially produced since the 1930s, volatilize into the atmosphere, and subsequently partition into water that is in contact with the atmosphere. Concentrations of CFCs have varied over time, facilitating their use in age-dating. In this report, CFCs refer to the CFCs that commonly were used for age-dating purposes during the time period 1992–2007: CFC-11 ($CFCl_3$), CFC-12 (CF_2Cl_2), and CFC-113 ($C_2F_3Cl_3$).

Contaminated   Where used in the context of CFC or SF_6 dating, indicates concentrations greater than those that would be found in groundwater that was at equilibrium with peak atmospheric CFC or SF_6 concentrations at the assumed recharge elevation and temperature.

Data Warehouse (DWH)   The DWH is a publically available database of NAWQA data http://infotrek.er.usgs.gov/traverse/f?p=NAWQA:HOME:0:.

Environmental Tracer   For the purposes of this report, an environmental tracer (or "tracer") is considered to be a widespread element, compound or isotope that is used to infer groundwater time-of-travel. The term "environmental" indicates widespread occurrence, as compared with a local-spatial-scale tracer injection used to understand tracer movement at one particular study site. In practice, the tracer (not the water) is dated.

Excess Air   Atmospheric air (gases), beyond the amount that can be attributed to air/water solubility, that is incorporated into shallow groundwater during or following recharge. Excess air is added to groundwater by air entrainment during infiltration and (or) by water-table fluctuations.

Land-Use Study (LUS)   A focused investigation of water-quality conditions associated with an individual land-use setting. LUSs generally are composed of shallow wells (usually monitoring wells) in recharge areas of regionally important land-use settings. Variability from hydrogeologic factors is reduced by restricting an individual LUS to a single aquifer.

Major-Aquifer Study (MAS)   A broad assessment of water-quality conditions in groundwater of a Study Unit, generally focusing on the shallow used resource. MASs typically are composed of networks of domestic supply wells located in principal aquifers. Major-Aquifer Studies were formerly called Study Unit Surveys.

Major Dissolved Gases N_2, O_2, Ar, CH_4, and CO_2. Four of these gases, N_2, O_2, Ar, and CO_2, are the volumetrically dominant gases in dry atmosphere that are incorporated into groundwater during recharge. Non-atmospheric CO_2 is incorporated into groundwater (for example, from root respiration and from redox reactions), generally at concentrations considerably greater than those from atmospheric sources. CH_4 can be incorporated into groundwater as a result of redox reactions; although CH_4 is not ubiquitously detected in groundwater, CH_4 concentrations often become a volumetrically major dissolved gas. These five gases, often analyzed as a suite, can be used to infer recharge temperatures (N_2, Ar), redox state (N_2, O_2, CH_4), and excess N_2 from denitrification.

Modern   Where used in the context of tracer-based CFC-based piston-flow ages, slight enrichment above concentrations that would be found in groundwater that was at equilibrium with peak atmospheric CFC concentrations may be referred to as "modern", to differentiate minor enrichment from greater enrichment. Minor enrichment is enrichment that could reasonably be attributed to uncertainties in the methodology, for example: young groundwater recharged under conditions of a local atmospheric CFC anomaly, or minor bias in estimated recharge temperature or recharge elevation. Greater enrichment frequently indicates the presence of CFCs from non-atmospheric sources. This term is used in this report only in appendix A.

Piston Flow   A simplified and idealized concept of groundwater flow in which groundwater moves in discrete packets by advection only, without hydrodynamic dispersion or mixing.

Reference (REF) Well   A well installed with the intent of providing samples that represent groundwater that has been minimally impacted by anthropogenic activities, that is, background or natural groundwater.

SF₆ SF_6, or sulfur hexafluoride, is a compound that has both natural and anthropogenic sources. Like CFCs, SF_6 volatilizes into the atmosphere and subsequently partitions into water that is in contact with the atmosphere. SF_6 concentrations have been increasing over time, facilitating the use of SF_6 in age-dating applications where natural sources of SF_6 are absent or negligible.

Study Unit   A major hydrologic system of the United States in which the NAWQA Program has focused water-quality studies. Study Units are combinations of ground- and surface-water systems. Most NAWQA Study Units are greater than 4,000 square miles (10,000 square kilometers) in area.

Study-Unit Survey (SUS)   See Major-Aquifer Study (MAS).

Tracer-based piston-flow age   The time-of-travel of groundwater, assuming that piston-flow conditions and assuming the time-of-travel implied by the tracers reflects the time-of-travel of the groundwater.

Table 1   11

Table 1.   Summary of tracer-based piston-flow ages that were interpreted in this report.

Table 1 provides a summary of the tracer-based piston-flow ages by sample. Data are available for download as a Microsoft© Excel file at http://pubs.usgs.gov/sir/2012/5141/.

Appendix A. Summary of CFC, SF$_6$, and ^{3}H/^{3}He Tracer Samples Collected from Selected NAWQA Networks between Fiscal Years 2006 and 2010

CFC, SF$_6$, and ^{3}H-^{3}He tracer data collected from NAWQA LUS, MAS, FPS, and REF networks between fiscal years 2006 and 2010 are documented in table A1. Data are available for download as a Microsoft© Excel file at http://pubs.usgs.gov/sir/2012/5141/.

Appendix B. Documentation and Analysis of Interpreted CFC, SF$_6$, and ^{3}H/^{3}He Data

Tracer data from a total of 812 sites are summarized in table 1 in the body of the report. The tracer data interpretations for these sites, along with selected ancillary chemical and well construction data, are presented in this appendix and are organized by Study Unit and network. Tables mentioned in this appendix are available for download as Microsoft© Excel files at http://pubs.usgs.gov/sir/2012/5141/.

A synopsis is given for each network, which includes:

- Number of wells in each network,

- Year of sample collection,

- Tracer(s) that were analyzed in the samples,

- General type of aquifer materials,

- Relevant notes about the approach used for estimating recharge temperature and (for CFCs and SF$_6$) excess air,

- A list of some of the notable advantages and disadvantages associated with the datasets,

- Range of water-level depths,

- A plot of tracer-based piston-flow age versus depth below water table (age gradient), (a reconstructed ^{3}H plot (where ^{3}H data were available), and

- A tracer-tracer comparison where possible.

The data also are presented in tables B1-B57 that include the raw tracer data, major dissolved-gas data, and the ancillary chemical and well construction data that are used in the interpretations.

The three types of figures used in this appendix (age gradient plots, reconstructed ^{3}H plots, and tracer age comparisons) can provide insight into tracer interpretations and the hydrologic systems associated with the tracer results. This was addressed in detail in Hinkle and others (2010) and will not be repeated in this discussion because the same procedures were used in the tracer interpretations.

Study Unit abbreviations in this appendix are defined in table A1 (appendix A). Network names in this appendix are as listed in the NAWQA DWH (http://infotrek.er.usgs.gov/traverse/f?p=NAWQA:HOME:0:). Well construction, water level, and ancillary chemistry data are from the NAWQA DWH.

Reference Cited

Hinkle, S.R., Shapiro, S.D., Plummer, L.N., Busenberg, E., Widman, P.K., Casile, G.C., and Wayland, J.E., 2010, Estimates of tracer-based piston-flow ages of groundwater from selected sites—National Water-Quality Assessment Program, 1992–2005: U.S. Geological Survey Scientific Investigations Report 2010-5229, 90 p. (Also available at http://pubs.usgs.gov/sir/2010/5229/.)

ACAD LUSRC1

Samples from seven sites in the ACAD Study Unit were collected during 2007 for CFCs and SF_6 (network and, in parentheses, number of sites):

.   LUSRC1 (7)

.   The wells were installed in the coastal lowlands aquifer system of the Chicot aquifer, Upper sand unit.

.   Major dissolved-gas data were available for all seven sites. Of these seven sites, only one was oxic and one site was degassed.

Age interpretations from tracer concentrations were made assuming that recharge elevation was equal to the elevation of the water table, that recharge temperature was equal to the mean annual air temperature +1°C, and that excess air concentrations were 2 cc STP/kg.

The raw tracer data, major dissolved-gas data, the ancillary chemical and well construction data that were used in the interpretations, and the piston-flow ages are presented in table B1.

.   Advantages associated with these samples:

.   Short open intervals (10 feet).

.   Median penetration of center of open interval into water table was 14.37 feet (sampling close to the water table, potentially minimizing mixing).

.   Monitoring wells, therefore low pumping stress.

.   Multiple tracers (CFCs and SF_6, as well as major dissolved gases).

.   Disadvantages associated with these samples:

.   Most sites were suboxic and one site degassed, so major dissolved-gas data not useful for defining recharge temperatures and excess air.

.   CFC data affected by degradation due to suboxic conditions.

.   Large discrepancy in recharge temperature estimates from major dissolved-gas data and MAAT +1°C.

.   Depth to water (can affect tracer transport to water table):

.   Median: 49.48 feet

.   Mean: 49.92 feet

.   Min: 41.51 feet

.   Max: 63.14 feet

.   Brief analysis:

.   The SF_6-based age gradient for these sites is shown in figure B1. There is no increase in age with depth. LUS wells generally are designed to be located in recharge areas; the large range in tracer ages at similar depths,

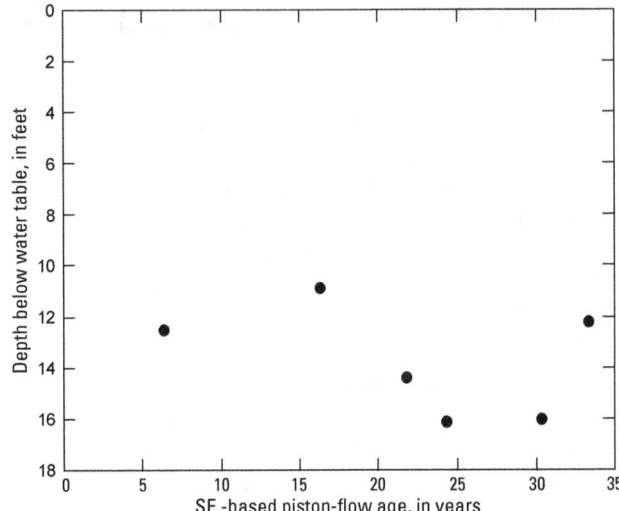

Figure B1. SF_6-based age gradient for dated sites from the LUSRC1 network, ACAD Study Unit.

however, indicates that these wells may be located in both recharge and discharge areas. Alternatively, the samples with the oldest SF_6-based ages correspond to the most degraded CFC samples, as well as samples that had methane present, indicating that degassing issues, such as gas stripping, may have altered the SF_6 concentrations leading to the older SF_6-based ages.

ACAD SUS1

Samples from the ACAD Study Unit were collected during 2007 and 2008 for CFCs, SF_6, and $^3H/^3He$ (network and, in parentheses, number of sites):

.   SUS1 (CFCs, 7 in 2007; SF_6, 7 in 2007, $^3H/^3He$, 19 in 2008).

The wells were installed in the coastal lowlands aquifer system of the Chicot aquifer, Upper sand unit, and the 20-foot sands of the Lake Charles Area.

Major dissolved-gas data were available for 7 sites in 2007 and 18 sites in 2008. Of the seven sites from 2007, three were oxic and four were suboxic. Of the 18 sites in 2008, only 1 site was oxic and 17 sites were suboxic.

Age interpretations from tracer concentrations were made assuming that recharge elevation was equal to the elevation of the water table, that recharge temperature was equal to the mean annual air temperature +1°C, and that excess air concentrations were 2 cc STP/kg.

$^3H/^3He$ ages were calculated for only 6 sites (only 1 of the 6 sites required a correction for terrigenic He), while 13 sites were not datable due to low tritium concentrations.

The raw tracer data, major dissolved-gas data, the ancillary chemical and well construction data that were used in the interpretations, and the piston-flow ages are presented in table B2.

. Advantages associated with these samples:

. Multiple tracers (CFCs, SF_6, and $^3H/^3He$, as well as major dissolved gases).

. Disadvantages associated with these samples:

. Open intervals of wells range from 5 to 30 feet.

. Median penetration of center of open interval into water table was 85.61 feet (not sampling close to the water table).

. Domestic wells.

. Generally highly reducing conditions. Only four sites contained >1 mg O_2/L (field O_2). Of the 25 samples with major dissolved-gas data, 18 contained detectable CH_4.

. CFC data affected by degradation due to suboxic conditions.

. Large discrepancy in recharge temperature estimates from major dissolved-gas data and MAAT +1°C.

. Depth to water (can affect tracer transport to water table):

. Median: 51.00 feet

. Mean: 50.85 feet

. Min: 4.33 feet

. Max: 91.01 feet

. Brief analysis:

. The SF_6- and $^3H/^3He$-based age gradients for these sites are shown in figures B2 and B3. SUS wells generally are randomly distributed wells (usually domestic supply wells), and the SUS wells in this ACAD network are domestic wells and generally are fairly deep. Despite the randomness of the distribution of the wells, there is an age structure with generally increasing tracer-based piston-flow age with increasing depth below the water table. Differences in screen length, recharge source/strength, aquifer heterogeneity, pumping stresses, and the position of the well within the flow system may cause some wells to deviate from the general pattern of increasing age with depth.

The reconstructed ^3H plots for CFC-, SF_6-, and $^3H/^3He$-based ages are shown in figures B4, B5, and B6. With only two datable CFC samples, and four datable SF_6 samples, there is limited value in the reconstructions. The $^3H/^3He$-based reconstruction, however, shows at least four samples that appear to be relatively unmixed and represent piston-flow transport, while two samples appear to be affected by mixing processes or diffusive ^3He loss. In addition, many of the $^3H/^3He$ samples that were not datable due to low tritium concentrations are still useful in determining that the water in these locations is old.

The SF_6- versus CFC-based age comparison for this network is shown in figure B7. The three sites sample oxic water and are not affected by suboxic conditions that would cause CFC-degradation. The age comparison for the three sites is good, but is complicated by the fact that the CFC-based ages appear to extend beyond the range of SF_6-based dating capability.

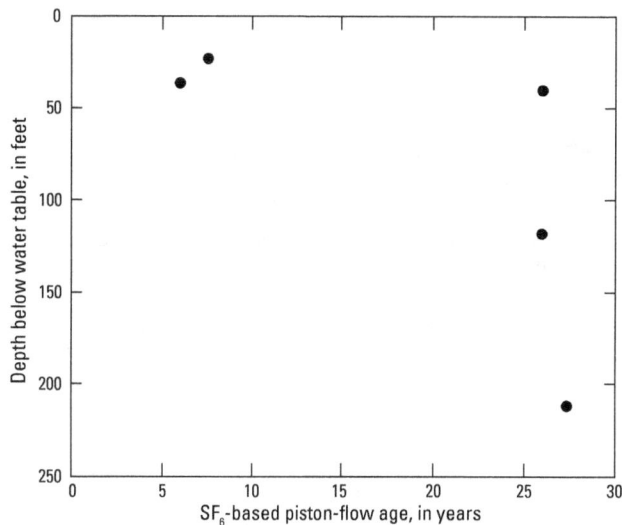

Figure B2. SF_6-based age gradient for dated sites from the SUS1 network, ACAD Study Unit.

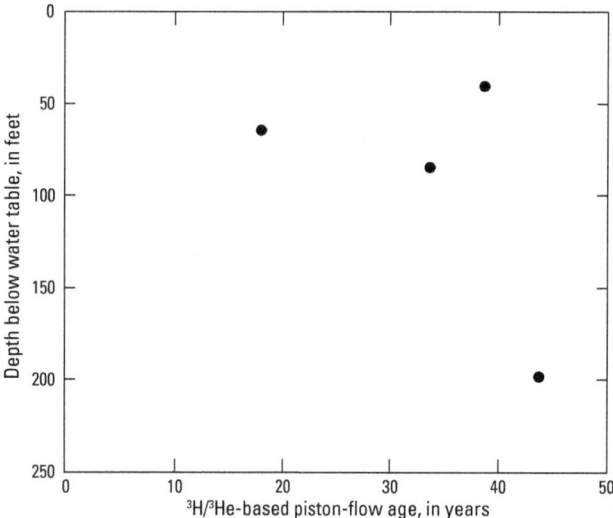

Figure B3. $^3H/^3He$-based age gradient for dated sites from the SUS1 network, ACAD Study Unit.

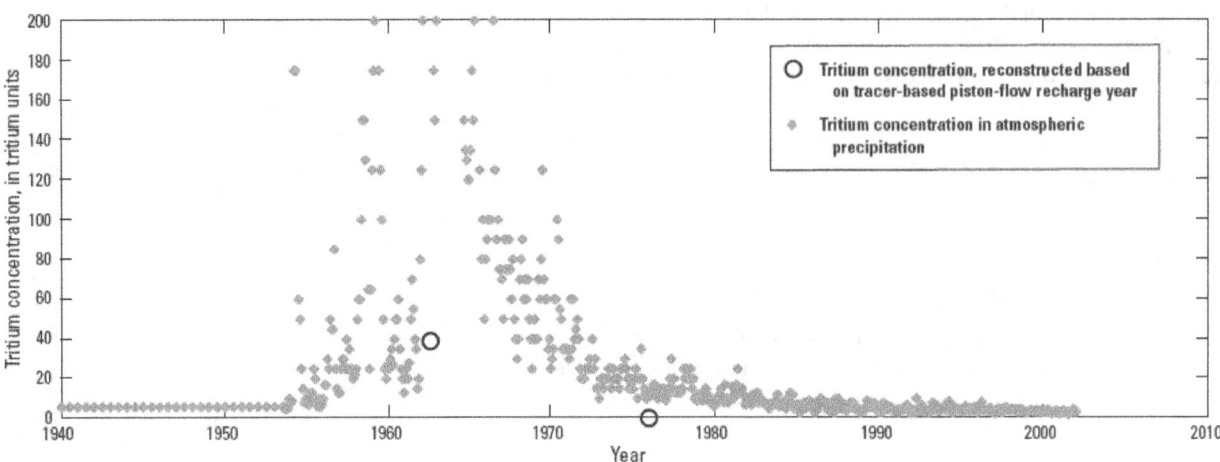

Figure B4.   Reconstructed tritium concentrations (using CFC-based ages) and tritium in atmospheric precipitation, SUS1 network, ACAD Study Unit.

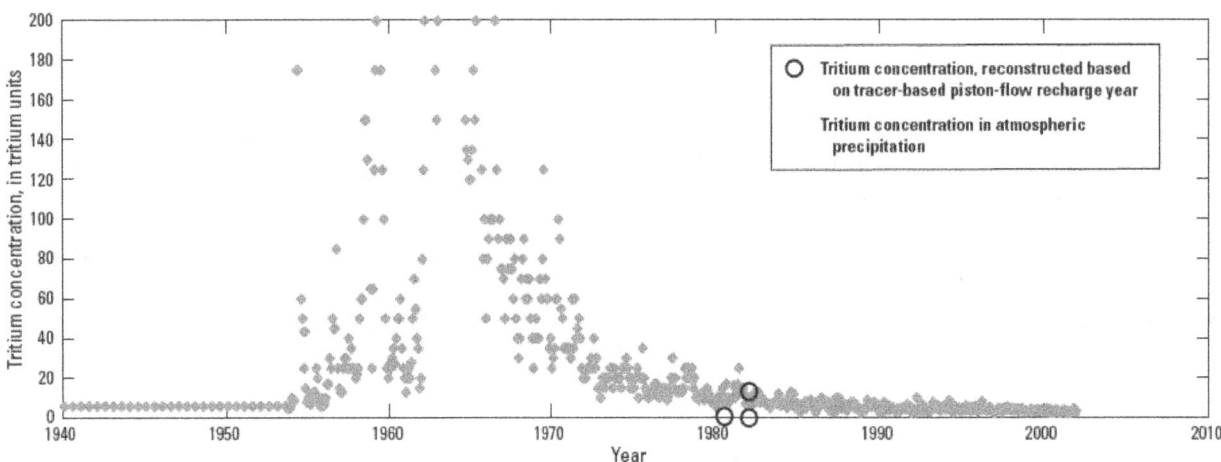

Figure B5.   Reconstructed tritium concentrations (using SF_6-based ages) and tritium in atmospheric precipitation, SUS1 network, ACAD Study Unit.

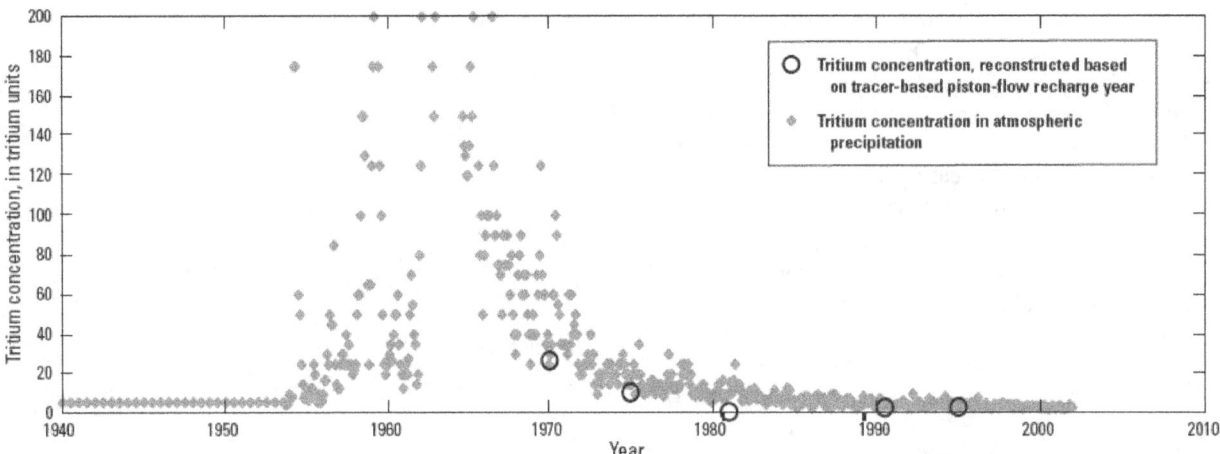

Figure B6.   Reconstructed tritium concentrations (using $^3H/^3He$-based ages) and tritium in atmospheric precipitation, SUS1 network, ACAD Study Unit.

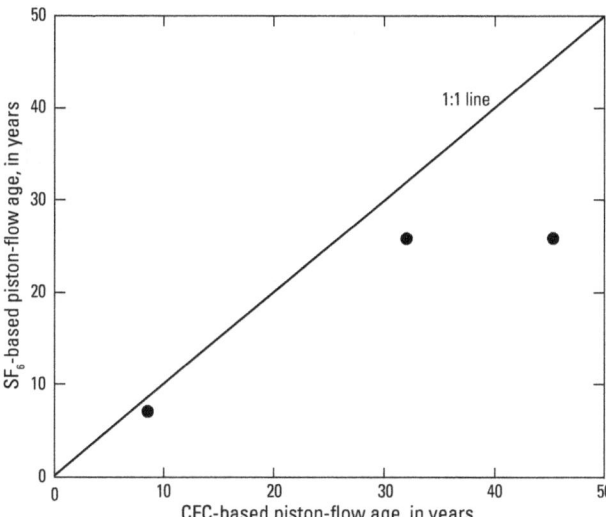

Figure B7. SF_6- versus CFC-based age comparison, SUS1 network, ACAD Study Unit.

ACFB SUS1, LUSCR3, and REFF04

Samples from 12 sites in the ACFB Study Unit were collected in 2007 for CFCs and $^3H/^3He$ (networks and, in parentheses, number of sites):

.   SUS1 (5)

.   LUSCR3 (5)

.   REFF04 (2)

The aquifer is composed of limestone (of the Floridian Aquifer System) and residium (of the surficial aquifer).

Age interpretations from tracer concentrations were made assuming that recharge elevation was equal to the elevation of the water table, that recharge temperature was equal to the mean annual air temperature $+1°C$, and that excess air concentrations were 2 cc STP/kg.

$^3H/^3He$ ages were calculated for 10 sites (none of the sites required a correction for terrigenic He), while 2 sites were not datable due to fractionation.

The raw tracer data, the ancillary chemical and well construction data that were used in the interpretations, and the piston-flow ages are presented in table B3.

.   Advantages associated with these samples:

.   Multiple tracers (CFCs and $^3H/^3He$).

.   4 domestic wells, 1 institutional well, and 7 unused wells, so generally low pumping rates.

.   All oxic sites so no problems with CFC degradation.

.   Disadvantages associated with these samples:

.   No major dissolved-gas data.

.   Relatively large open intervals ranging from 3.62–63 feet.

.   Depth to water (can affect tracer transport to water table):

.   Median: 39.48 feet

.   Mean: 39.71 feet

.   Min: 0.82 feet

.   Max: 101.05 feet

.   Brief analysis:

.   The CFC- and $^3H/^3He$-based age gradients for these sites are shown in figures B8 and B9. The CFC-based and $^3H/^3He$-based age gradients are similar, with relatively old ages at shallow depths, but with age generally increasing with depth. Differences in screen length, recharge source/strength, aquifer heterogeneity, pumping stresses, and the position of the well within the flow system may cause some wells to deviate from the general pattern of increasing age with depth. The $^3H/^3He$ ages are younger than the CFC ages at shallow depths, which may indicate that there was helium loss near the water table.

The reconstructed ^3H plots for CFC- and $^3H/^3He$-based ages are shown in figures B10 and B11. Both reconstructions show that samples appear to be relatively unmixed and represent piston-flow transport.

The $^3H/^3He$- versus CFC-based age comparison for this network is shown in figure B12. The age comparison is good, which indicates that the tracer-based piston-flow ages for this network can be used with a good deal of confidence.

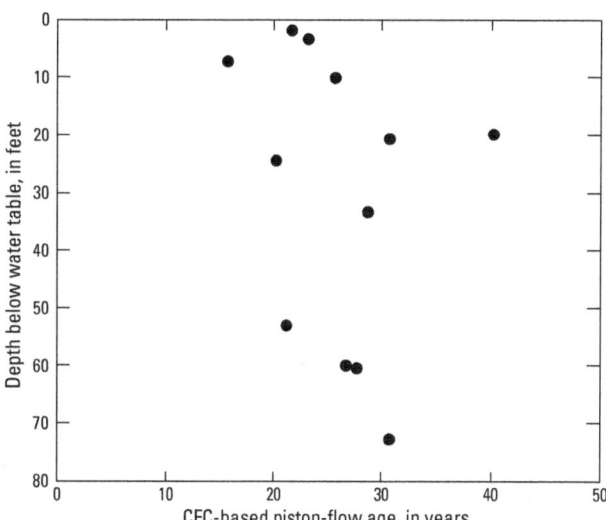

Figure B8.   CFC-based age gradient for dated sites from the SUS1, luscr3, and REFF04 networks, ACFB Study Unit.

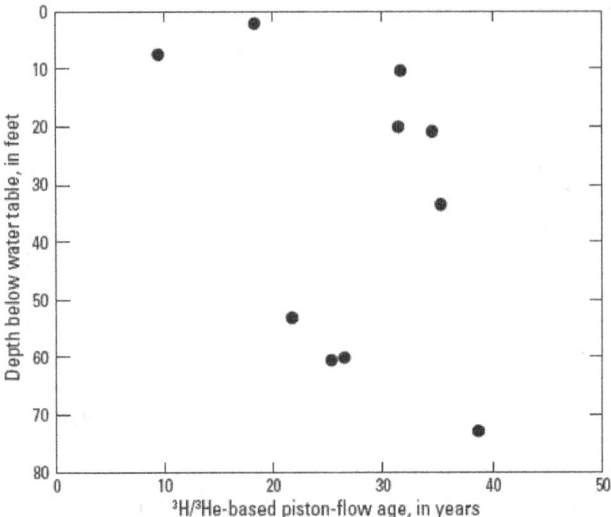

Figure B9. $^3H/^3He$-based age gradient for dated sites from the SUS1, LUSCR3, and REFF04 networks, ACFB Study Unit.

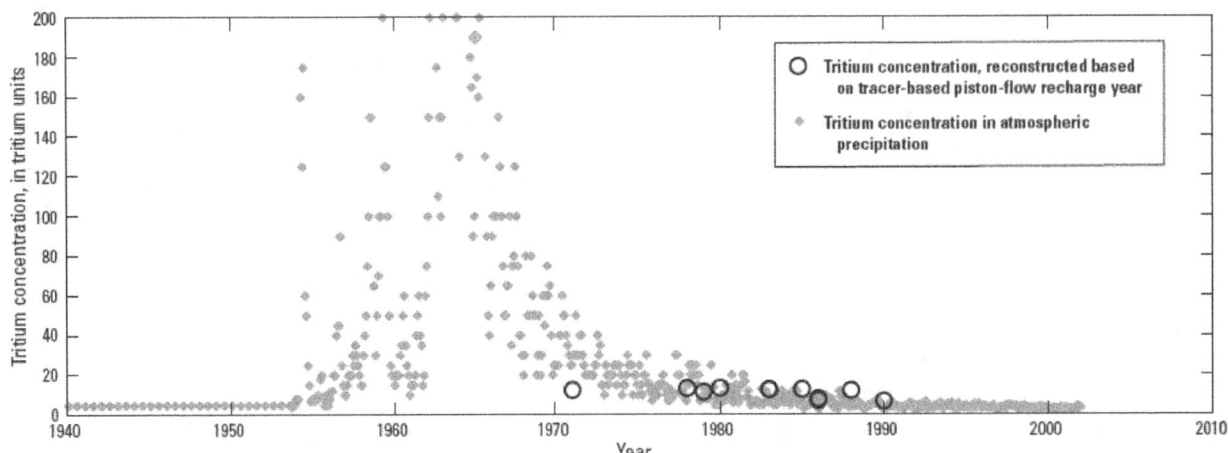

Figure B10. Reconstructed tritium concentrations (using CFC-based ages) and tritium in atmospheric precipitation, sus1, luscr3, and REFF04 networks, ACFB Study Unit.

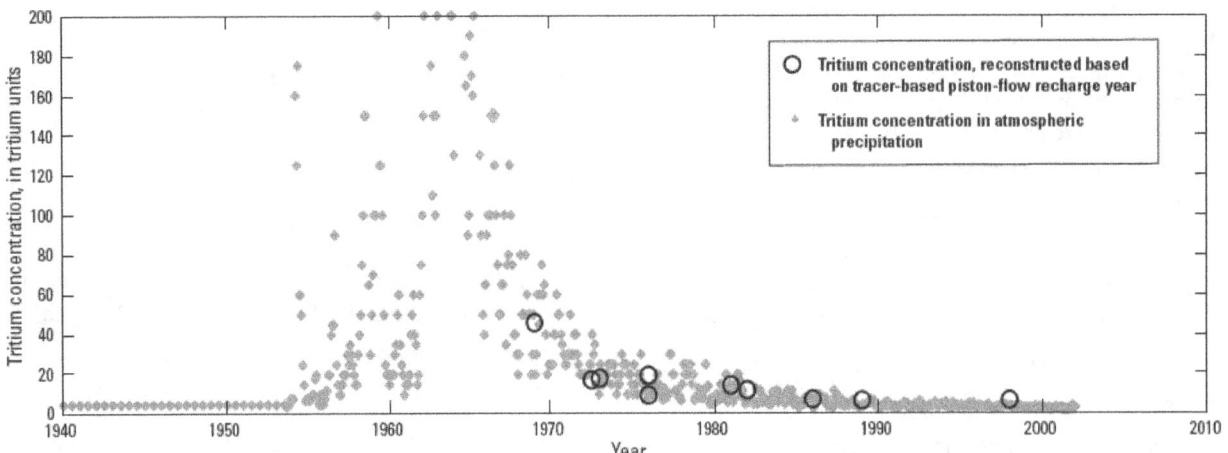

Figure B11. Reconstructed tritium concentrations (using $^3H/^3He$-based ages) and tritium in atmospheric precipitation, sus1, luscr3, and reff04 networks, ACFB Study Unit.

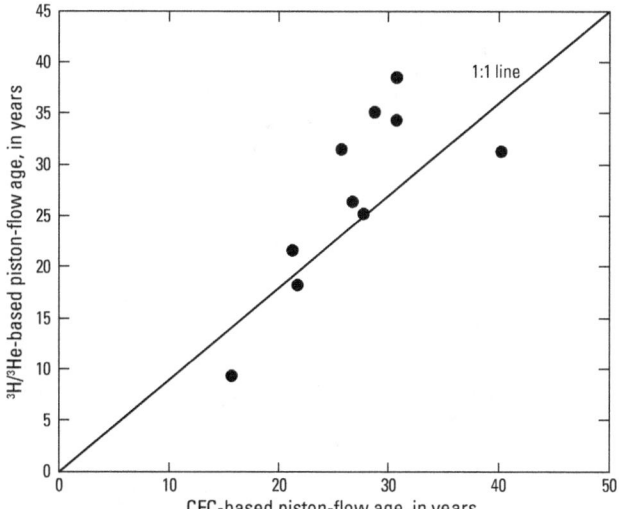

Figure B12. ^{3}H/^{3}He- versus CFC-based age comparison, SUS1, LUSCR3, AND REFF04 networks, ACFB Study Unit.

ALBE SUS7, LUSAG1, and REFFO1

Samples from 12 sites in the ALBE Study Unit were collected in 2007 for CFCs and ^{3}H/^{3}He (networks and, in parentheses, number of sites):

. SUS7 (5)

. LUSAG1 (5)

. REFFO1 (2)

The aquifer is composed of limestone (of the Castle Hayne Limestone) and sand and silt (of the surficial aquifer system).

Major dissolved-gas data were available for all 12 sites. Of these 12 sites, only 1 site was oxic and 11 were suboxic.

Age interpretations from tracer concentrations were made assuming that recharge elevation was equal to the elevation of the water table. Estimates of recharge temperature and excess air were based on major dissolved-gas data. Although most sites were suboxic, the dissolved-gas data did not show evidence of denitrification or other problems.

^{3}H/^{3}He ages were calculated for two sites (these two sites did not require a correction for terrigenic He), while five sites were not datable due to fractionation, one site was not datable due to low tritium, one site was not datable because there was no tritium measured, and samples from three sites were lost due to high pressure or other laboratory issues.

The raw tracer data, major dissolved-gas data, the ancillary chemical and well construction data that were used in the interpretations, and the piston-flow ages are presented in table B4.

. Advantages associated with these samples:

. Mostly unused wells drilled for water level or water quality observations, so generally low pumping rates.

. Multiple tracers (CFCs and ^{3}H/^{3}He, as well as major dissolved gases).

. Disadvantages associated with these samples:

. Only one oxic site, CFCs affected by degradation.

. Relatively large open intervals ranging from 1.87 to 33 feet so mixing likely.

. Median penetration of center of open interval into water table was 19.83 feet (sampling close to the water table, potentially minimizing mixing).

. Depth to water (can affect tracer transport to water table):

. Median: 6.67 feet

. Mean: 7.78 feet

. Min: 0.43 feet

. Max: 26.17 feet

. Brief analysis:

. The ^{3}H/^{3}He-based age gradient for these sites is shown in figure B13. The ^{3}H/^{3}He-based age gradient generally increases with depth, with very young ages near the surface and old ages at depth. Differences in screen length, recharge source/strength, aquifer heterogeneity, pumping stresses, and the position of the well within the flow system may cause some wells to deviate from the general pattern of increasing age with depth.

The reconstructed ^{3}H plots for CFC- and ^{3}H/^{3}He-based ages are shown in figures B14 and B15. There is only one sample in the CFC-based reconstruction, but it plots within the given range for the ^{3}H input function. In the ^{3}H/^{3}He-based reconstruction, three samples (1 LUSAG, 2 REF) appear to be relatively unmixed and represent piston-flow transport, while two samples (SUS wells) appear to have been affected by dispersion and/or mixing processes. The two SUS wells are deep wells with large open intervals that likely capture water with a large range of ages.

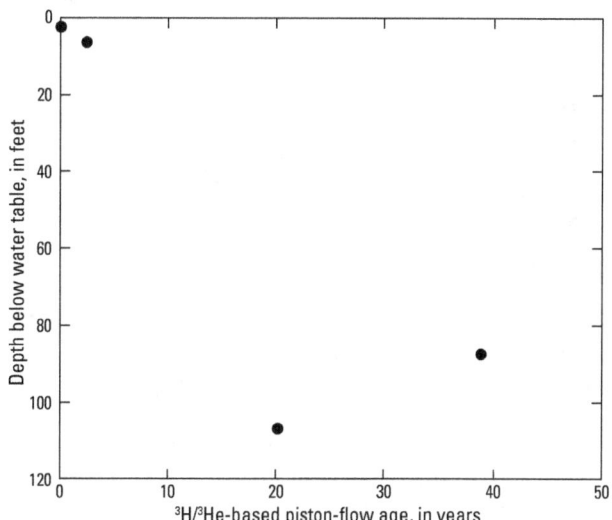

Figure B13. ^{3}H/^{3}He-based age gradient for dated sites from the sus7, lusag1, and REFFO1 networks, ALBE Study Unit.

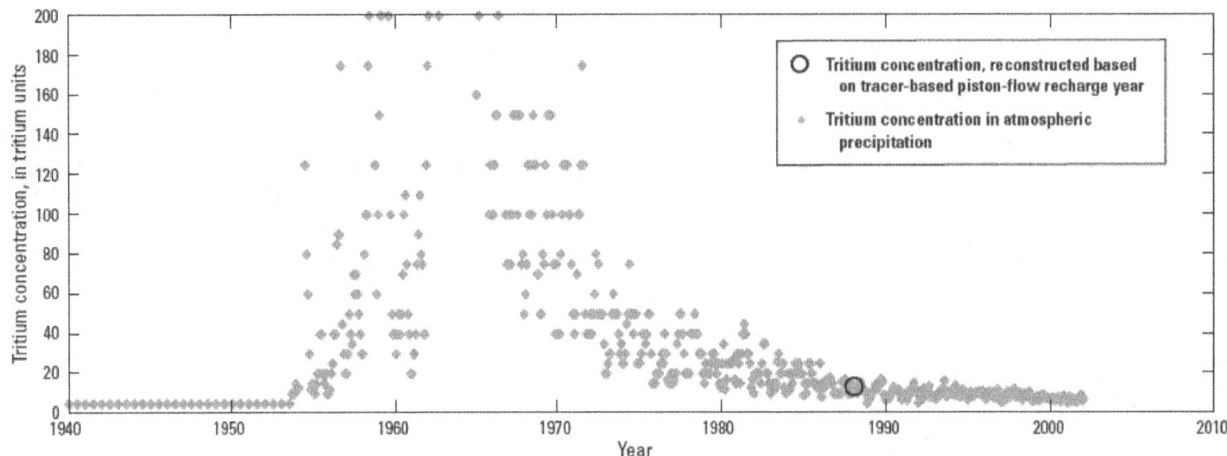

Figure B14.   Reconstructed tritium concentrations (using CFC-based ages) and tritium in atmospheric precipitation, SUS7, LUSAG1, AND REFF01 networks, ALBE Study Unit.

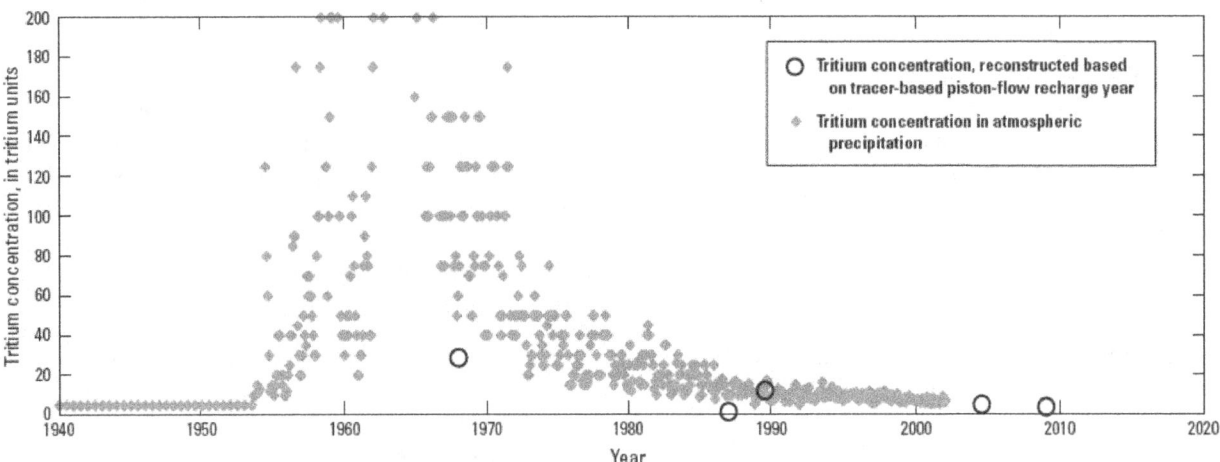

Figure B15.   Reconstructed tritium concentrations (using ^{3}H/^{3}He-based ages) and tritium in atmospheric precipitation, SUS7, LUSAG1, AND REFF01 networks, ALBE Study Unit.

ALBE SUS8

Samples from sites in the ALBE Study Unit were collected in 2007 for CFCs (17) and ^{3}H/^{3}He (5) (networks and, in parentheses, number of sites):

.   SUS8 (17)

The aquifer is composed of Piedmont and Blue Ridge crystalline-rock.

Major dissolved-gas data were available for 17 sites. Of these 17 sites, 9 were oxic and 8 were suboxic.

Age interpretations from tracer concentrations were made assuming that recharge elevation was equal to the elevation of the water table. Estimates of recharge temperature and excess air were based on major dissolved-gas data, with recharge temperature and excess air at suboxic sites being constrained using median excess air at oxic sites.

^{3}H/^{3}He ages were calculated for three sites (only one of the three sites required a correction for terrigenic helium), while one site was not datable due to fractionation, and the sample from one site was lost due to high pressure or other laboratory issues.

The raw tracer data, major dissolved-gas data, the ancillary chemical and well construction data that were used in the interpretations, and the piston-flow ages are presented in table B5.

.   Advantages associated with these samples:

.   Domestic wells, so generally low pumping rates.

.   Multiple tracers (CFCs and ^{3}H/^{3}He, as well as major dissolved gases).

.   Disadvantages associated with these samples:

.   Relatively large open intervals ranging from 5 to 354 feet so mixing likely.

Median penetration of center of open interval into water table was 105.9 feet (not sampling close to the water table, potentially mixing).

Depth to water (can affect tracer transport to water table):

Median: 34.61 feet

Mean: 37.62 feet

Min: 22.07 feet

Max: 64.9 feet

Brief analysis:

The age gradient for these sites is shown in figure B16. The data show a great deal of scatter, which would be expected from SUS wells with large open intervals.

The reconstructed ^3H plots for CFC- and $^3H/^3He$-based ages are shown in figures B17 and B18. Several samples plot to the right of the ^3H reconstruction, which might suggest that there is either some tritium enrichment or mixed waters. The remaining samples, however, are consistent with the ^3H input function.

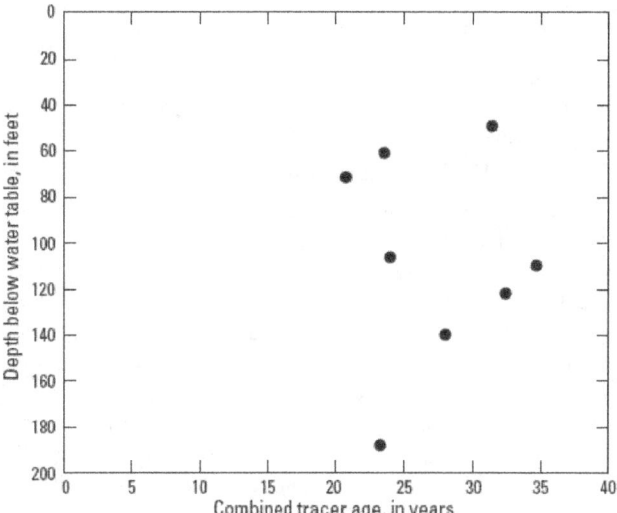

Figure B16. Age gradient for dated sites from the SUS8 network, ALBE Study Unit.

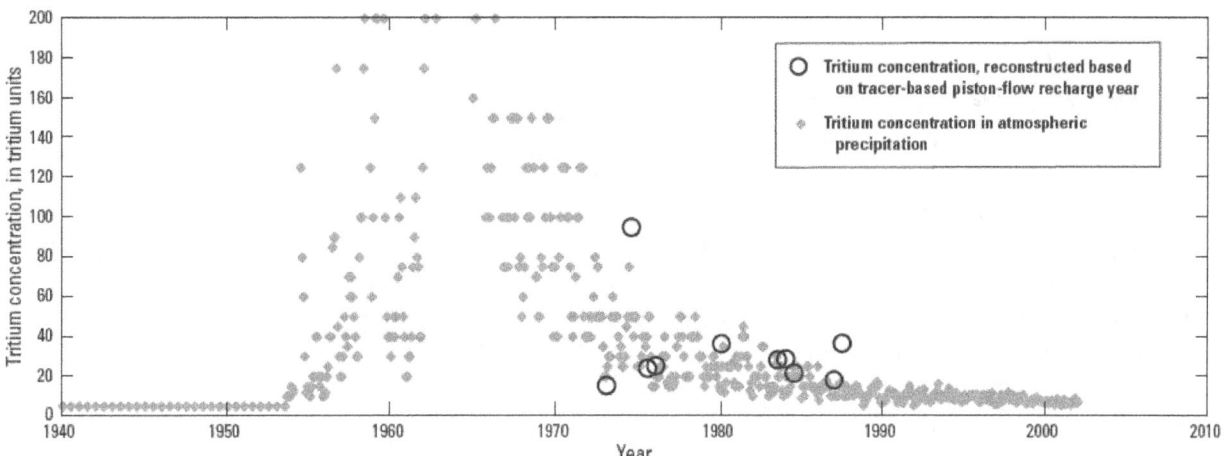

Figure B17. Reconstructed tritium concentrations (using CFC-based ages) and tritium in atmospheric precipitation, SUS8 network, ALBE Study Unit.

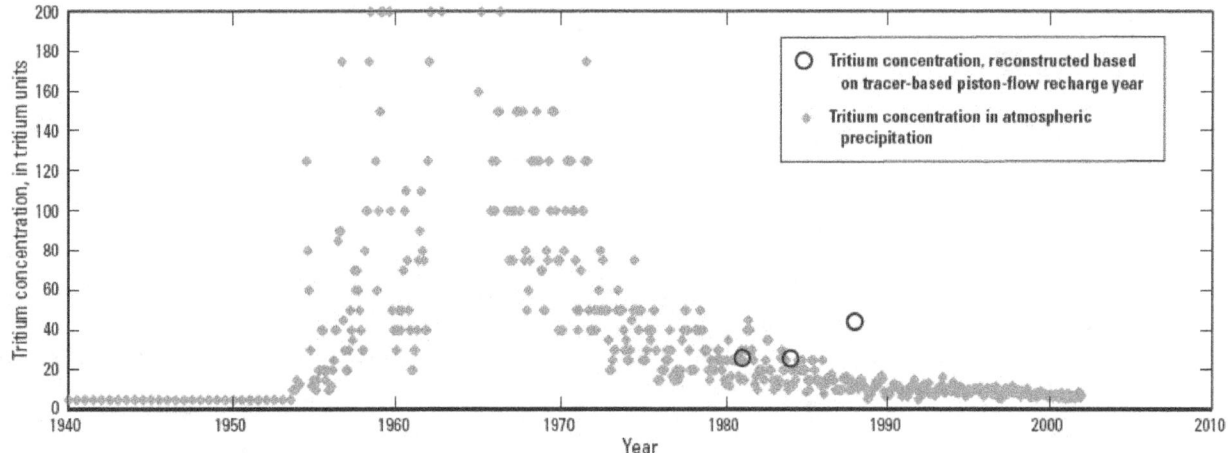

Figure B18. Reconstructed tritium concentrations (using $^3H/^3He$-based ages) and tritium in atmospheric precipitation, SUS8 network, ALBE Study Unit.

CAZB SUS1a

Samples from 8 sites in the CAZB Study Unit were collected in 2008 for ^{3}H/^{3}He (networks and, in parentheses, number of sites):

. SUS1a (^{3}H/^{3}He, 8; major dissolved gases, 35)

The aquifer is composed of Basin and Range basin-fill sands, gravel, clay, and conglomerates.

Major dissolved-gas data were available for 35 sites. Of these 35 sites, 29 were oxic and 6 were suboxic.

Age interpretations from tracer concentrations were made assuming that recharge elevation was equal to the elevation of the water table. Estimates of recharge temperature and excess air were based on major dissolved-gas data, with recharge temperature and excess air at suboxic sites being constrained using median excess air at oxic sites.

^{3}H/^{3}He ages were calculated for eight sites (only one of the eight sites required a correction for terrigenic helium).

The raw tracer data, major dissolved-gas data, the ancillary chemical and well construction data that were used in the interpretations, and the piston-flow ages are presented in table B6.

. Advantages associated with these samples:

. ^{3}H/^{3}He, as well as major dissolved gases.

. Disadvantages associated with these samples:

. Mixture of domestic, irrigation, stock, and public supply wells, so variable pumping rates.

. Relatively large open intervals ranging from 4 to 650 feet so mixing likely.

. Median penetration of center of open interval into water table was 212.9 feet (not sampling close to the water table, potentially mixing).

. Depth to water (can affect tracer transport to water table):

. Median: 208.61 feet

. Mean: 218.36 feet

. Min: 7.36 feet

. Max: 619.06 feet

. Brief analysis:

. The ^{3}H/^{3}He-based age gradient for these sites is shown in figure B19. Despite sampling from a mixture of well types, with relatively deep unsaturated zones and long open intervals, there is a structure to the age gradient with young ages at shallow depths and older ages at deeper intervals. Differences in screen length, recharge source/strength, aquifer heterogeneity, pumping stresses, and the position of the well within the flow system may cause some wells to deviate from the general pattern of increasing age with depth.

The reconstructed ^{3}H plot for CFC-based ages is shown in figure B20. The reconstruction shows that samples appear to be relatively unmixed and represent piston-flow conditions.

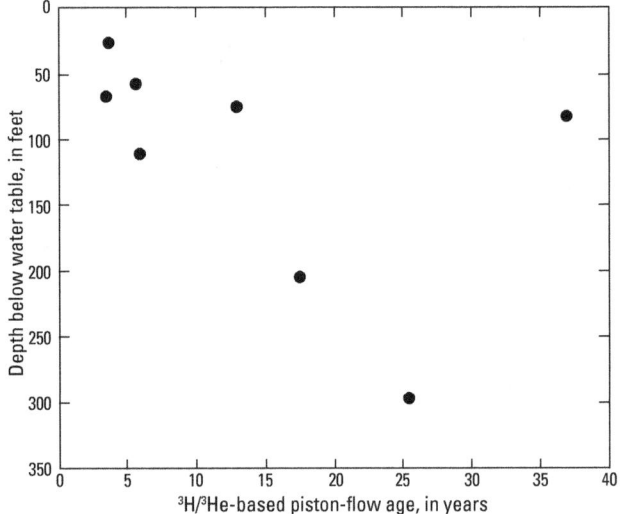

Figure B19. ^{3}H/^{3}He-based age gradient for dated sites from the SUS1a network, CAZB Study Unit.

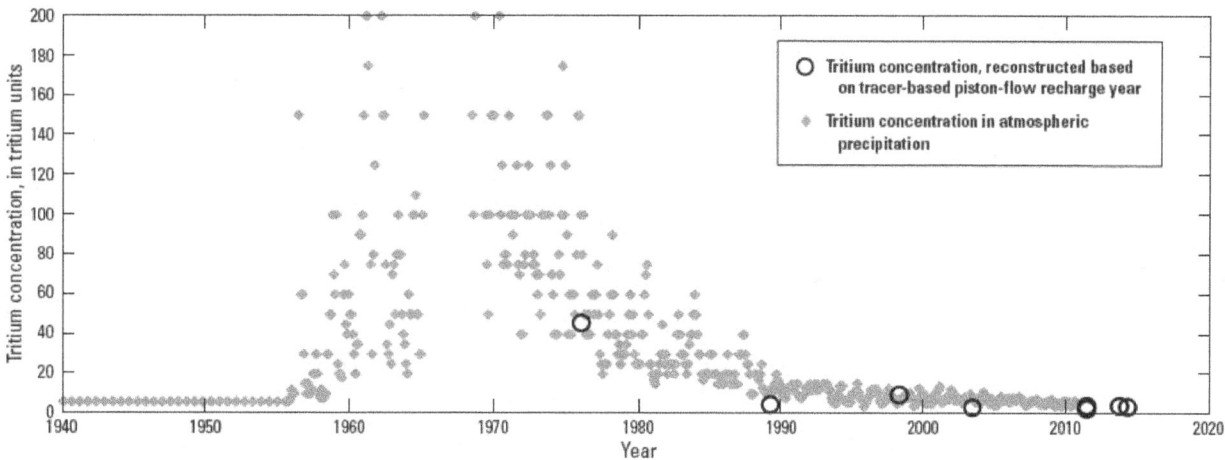

Figure B20. Reconstructed tritium concentrations (using $^3H/^3He$-based ages) and tritium in atmospheric precipitation, SUS1a network, CAZB Study Unit.

CCYK CCPTLUSAG2b

Samples from five sites in the CCYK Study Unit were collected in 2008 for CFCs, SF_6, and $^3H/^3He$ (networks and, in parentheses, number of sites):

. CCPTLUSAG2b (5)

The aquifer is composed of sand, silt, gravel, and clay.

Major dissolved-gas data were available for all five of the sites samples. Of these five sites, all five were oxic.

Age interpretations from tracer concentrations were made assuming that recharge elevation was equal to the elevation of the water table. Estimates of recharge temperature and excess air were based on major dissolved-gas data.

$^3H/^3He$ ages were calculated for three sites (no sites required a correction for terrigenic He), while two sites were not datable because of fractionation.

The raw tracer data, major dissolved-gas data, the ancillary chemical and well construction data that were used in the interpretations, and the piston-flow ages are presented in table B7.

. Advantages associated with these samples:

. Monitoring wells, so generally low pumping rates.

. Multiple tracers (CFCs, SF_6, and $^3H/^3He$, as well as major dissolved gases).

. Relatively small open intervals ranging from 5 to 15 feet so mixing minimized.

. Median penetration of center of open interval into water table was 14.91 feet (sampling close to the water table, potentially reduces mixing).

. Disadvantages associated with these samples:

. None.

. Depth to water (can affect tracer transport to water table):

. Median: 40.11 feet

. Mean: 33.4 feet

. Min: 4.45 feet

. Max: 66.86 feet

. Brief analysis:

The CFC-, SF_6-, and $^3H/^3He$-based age gradients for these sites are shown in figures B21, B22, and B23. All three gradients show a general structure of increasing age with depth, however, the $^3H/^3He$ ages are somewhat younger in the shallow wells indicating possible helium loss near the water table. Differences in screen length, recharge source/strength, aquifer heterogeneity, pumping stresses, and the position of the well within the flow system may cause some wells to deviate from the general pattern of increasing age with depth.

The reconstructed ^3H plots for CFC-, SF_6-, and $^3H/^3He$-based ages are shown in figures B24, B25, and B26. The three reconstructions are similar, and all of them have samples that plot above the ^3H input function.

The $^3H/^3He$- versus CFC-based age comparison and the SF_6- versus CFC-based age comparison for this network are shown in figures B27 and B28. The age comparisons show the same bias of younger $^3H/^3He$ ages, likely resulting from degassing near the water table, but comparable CFC- and SF_6-based ages.

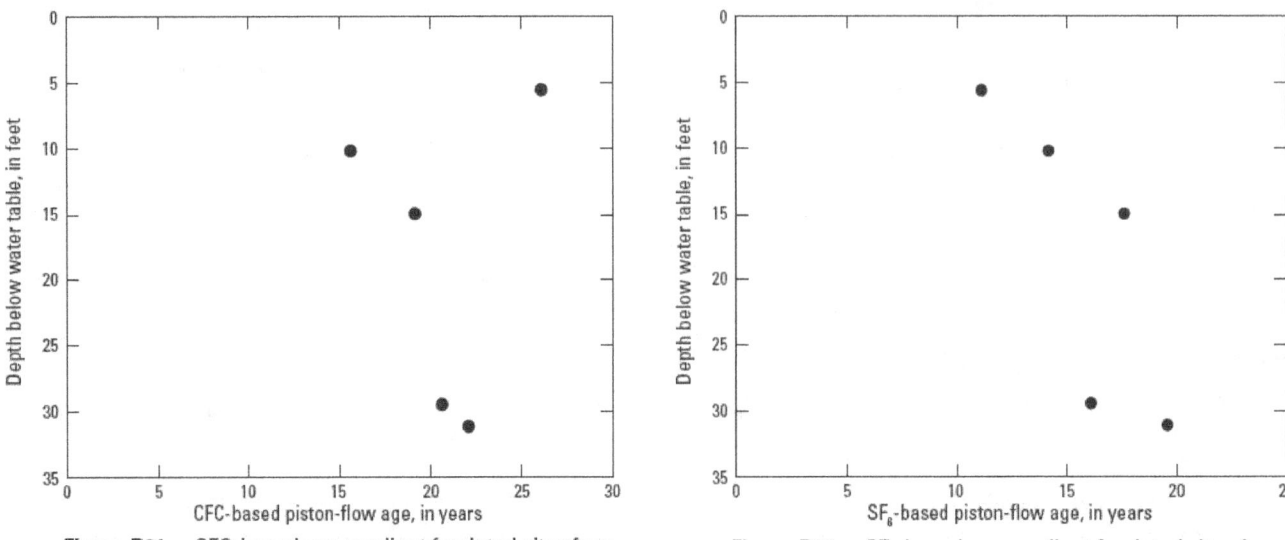

Figure B21. CFC-based age gradient for dated sites from the CCPTLUSAG2b network, CCYK Study Unit.

Figure B22. SF₆-based age gradient for dated sites from the CCPTLUSAG2b network, CCYK Study Unit.

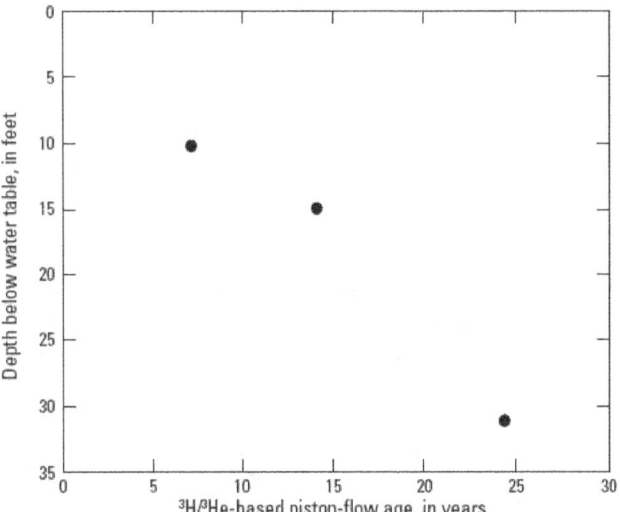

Figure B23. ³H/³He-based age gradient for dated sites from the CCPTLUSAG2b network, CCYK Study Unit.

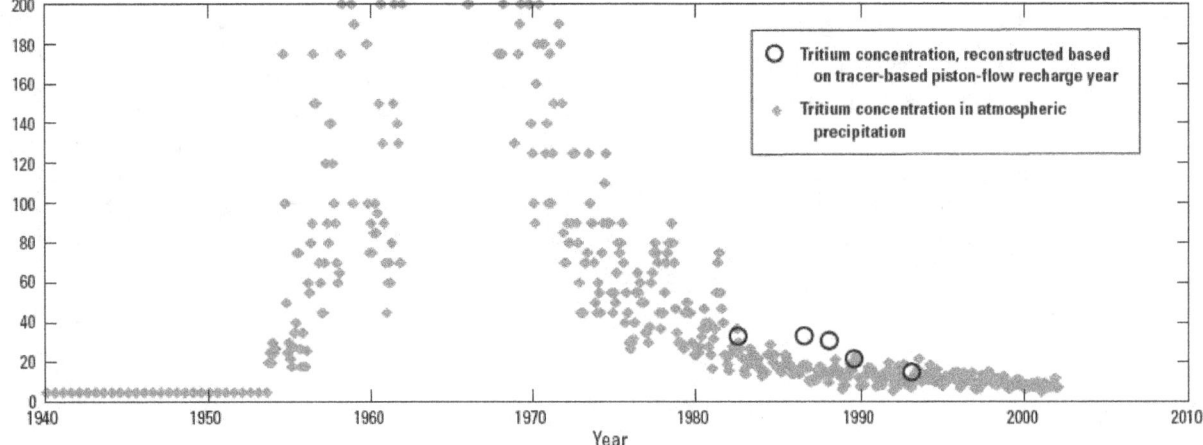

Figure B24. Reconstructed tritium concentrations (using CFC-based ages) and tritium in atmospheric precipitation, CCPTLUSAG2b network, CCYK Study Unit.

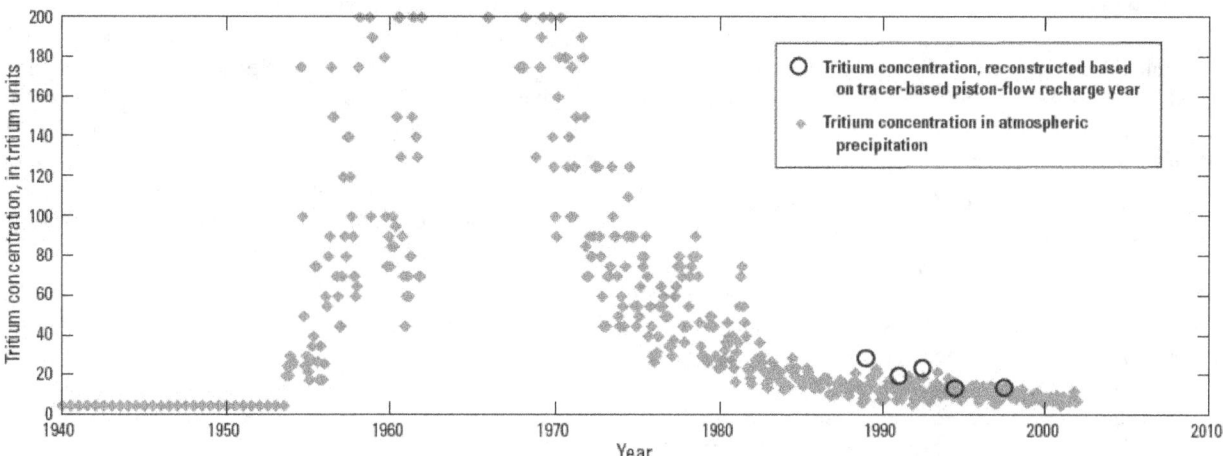

Figure B25. Reconstructed tritium concentrations (using SF$_6$-based ages) and tritium in atmospheric precipitation, CCPTLUSAG2b network, CCYK Study Unit.

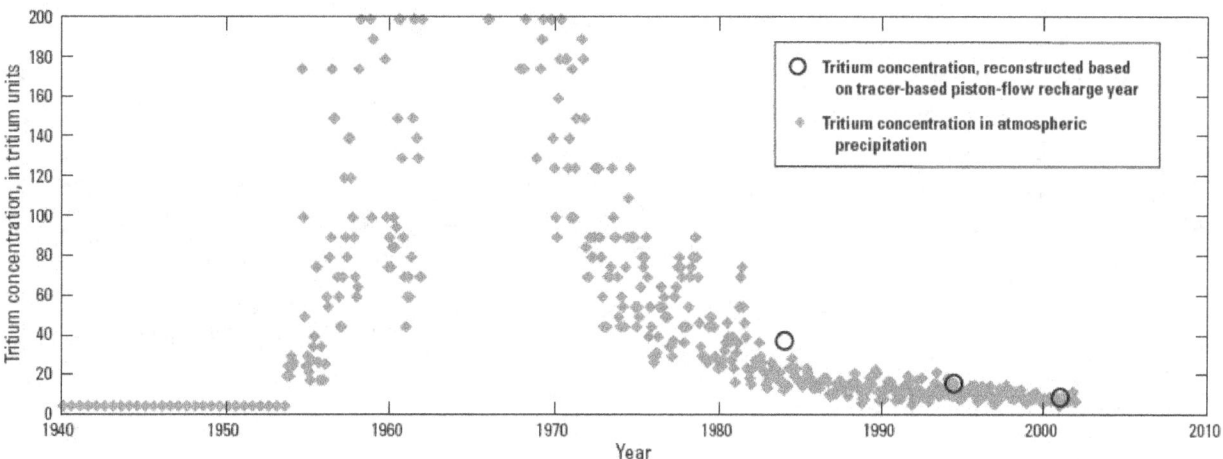

Figure B26. Reconstructed tritium concentrations (using ^{3}H/^{3}He-based ages) and tritium in atmospheric precipitation, CCPTLUSAG2b network, CCYK Study Unit.

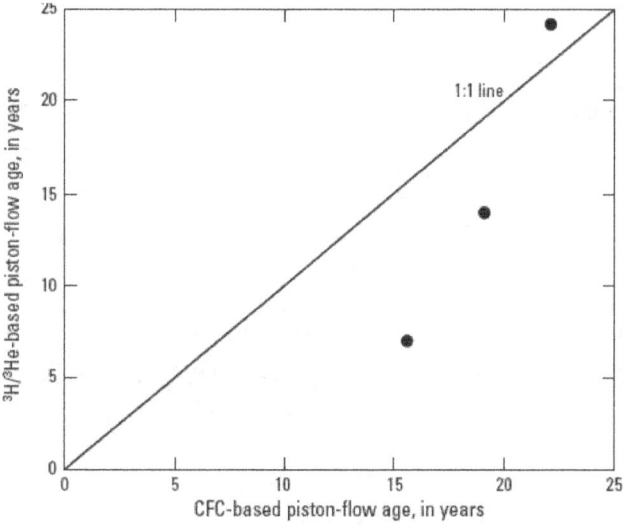

Figure B27. ^{3}H/^{3}He- versus CFC-based age comparison, CCPTLUSAG2b network, CCYK Study Unit.

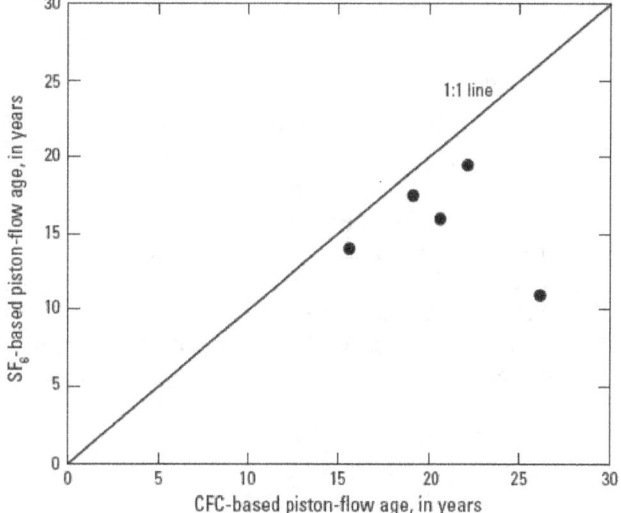

Figure B28. SF$_6$- versus CFC-based age comparison, CCPTLUSAG2b network, CCYK Study Unit.

CCYK CCPTLUSOR1b

Samples from five sites in the CCYK Study Unit were collected in 2008 for CFCs, SF_6, and $^3H/^3He$ (networks and, in parentheses, number of sites):

.   CCPTLUSOR1b (5)

The aquifer is composed of overburden sands and silts.

Major dissolved-gas data were available for all five sites. Of these five sites, all five were oxic.

Age interpretations from tracer concentrations were made assuming that recharge elevation was equal to the elevation of the water table. Estimates of recharge temperature and excess air were based on major dissolved-gas data.

$^3H/^3He$ ages were calculated for two sites (none of these sites required a correction for terrigenic He), while three sites were not datable due to fractionation.

The raw tracer data, major dissolved-gas data, the ancillary chemical and well construction data that were used in the interpretations, and the piston-flow ages are presented in table B8.

.   Advantages associated with these samples:

.   Multiple tracers (CFCs, SF_6, and $^3H/^3He$, as well as major dissolved gases).

.   Monitoring wells so generally low pumping stress.

.   Relatively short open intervals, ranging from 5 to 5.5 feet.

.   Median penetration of center of open interval into water table was 13.26 feet (sampling close to the water table, potentially minimizes mixing).

.   Disadvantages associated with these samples:

.   None.

.   Depth to water (can affect tracer transport to water table):

.   Median: 15.59 feet

.   Mean: 19.68 feet

.   Min: 5.03 feet

.   Max: 46.71 feet

.   Brief analysis:

.   The CFC- and SF_6-based age gradients for these sites are shown in figures B29 and B30. Both gradients show a general structure of increasing age with depth, and relatively old ages at shallow depths. Differences in screen length, recharge source/strength, aquifer heterogeneity, pumping stresses, and the position of the well within the flow system may cause some wells to deviate from the general pattern of increasing age with depth.

The reconstructed ^3H plots for CFC-, SF_6-, and $^3H/^3He$-based ages are shown in figures B31, B32, and B33. All three reconstructions show a general consistency with unmixed, piston-flow transport, however, the CFC- and SF_6-based reconstructions have samples that plot above the ^3H input function.

The SF_6- versus CFC-based age comparison for this network is shown in figure B34. The age comparison shows a great deal of scatter with no particular bias.

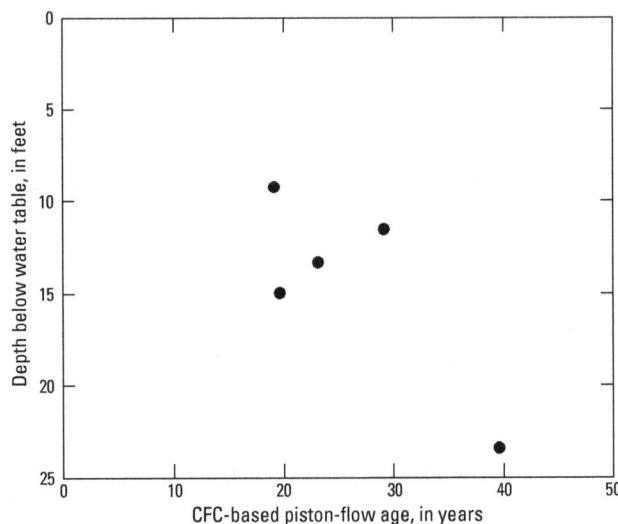

Figure B29.   CFC-based age gradient for dated sites from the CCPTLUSOR1b network, CCYK Study Unit.

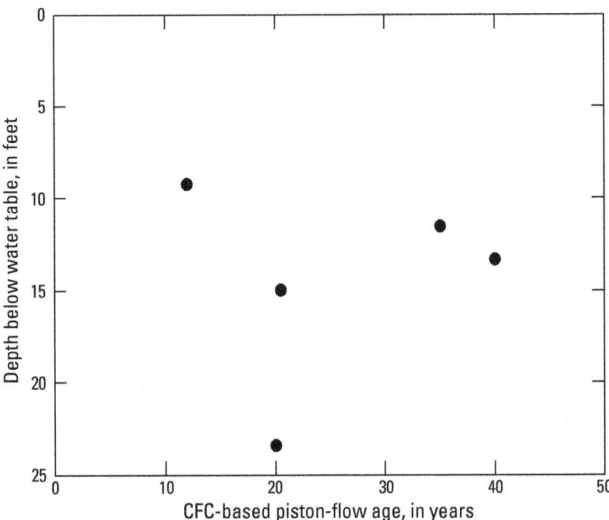

Figure B30. SF_6-based age gradient for dated sites from the CCPTLUSOR1b network, CCYK Study Unit.

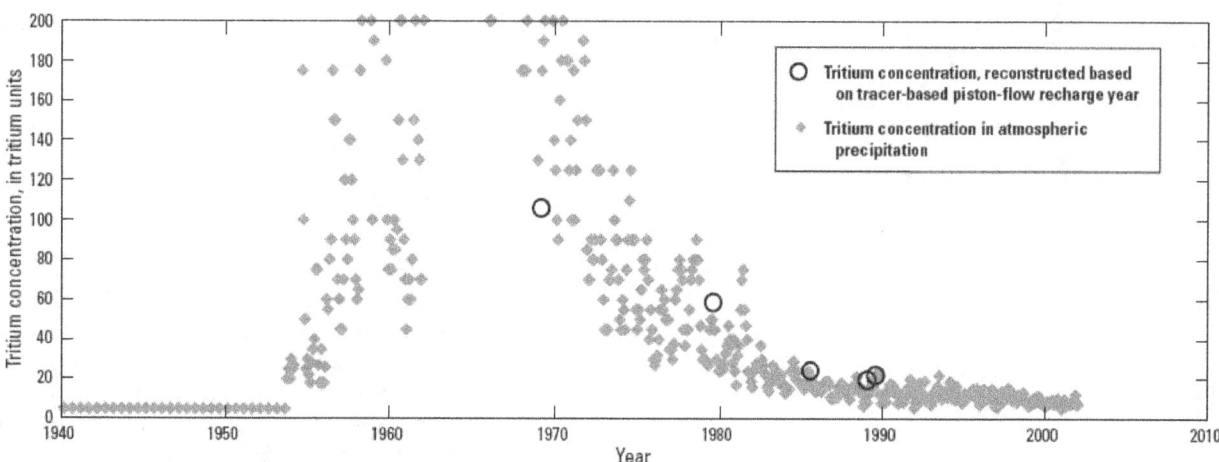

Figure B31. Reconstructed tritium concentrations (using CFC-based ages) and tritium in atmospheric precipitation, CCPTLUSOR1b network, CCYK Study Unit.

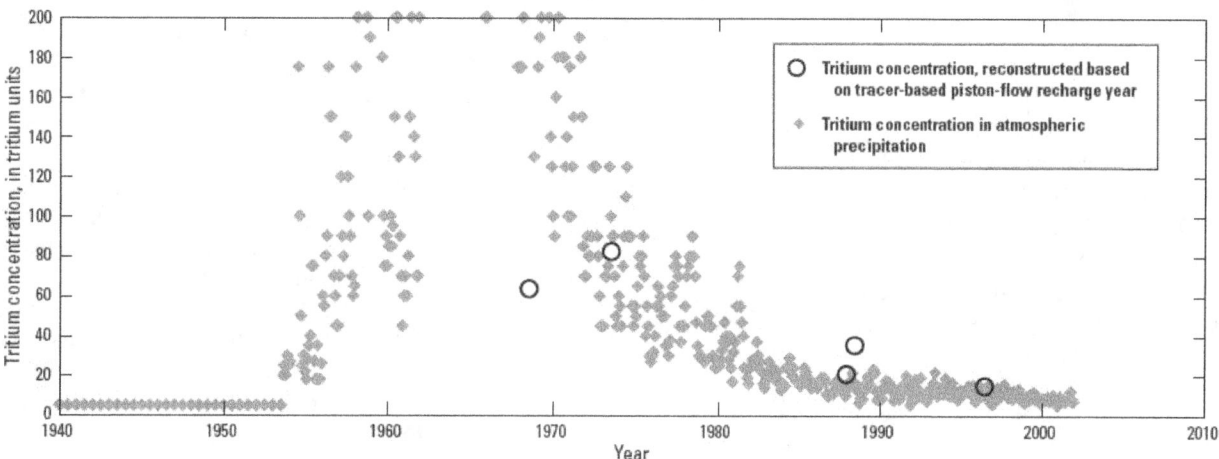

Figure B32. Reconstructed tritium concentrations (using SF_6-based ages) and tritium in atmospheric precipitation, CCPTLUSOR1b network, CCYK Study Unit.

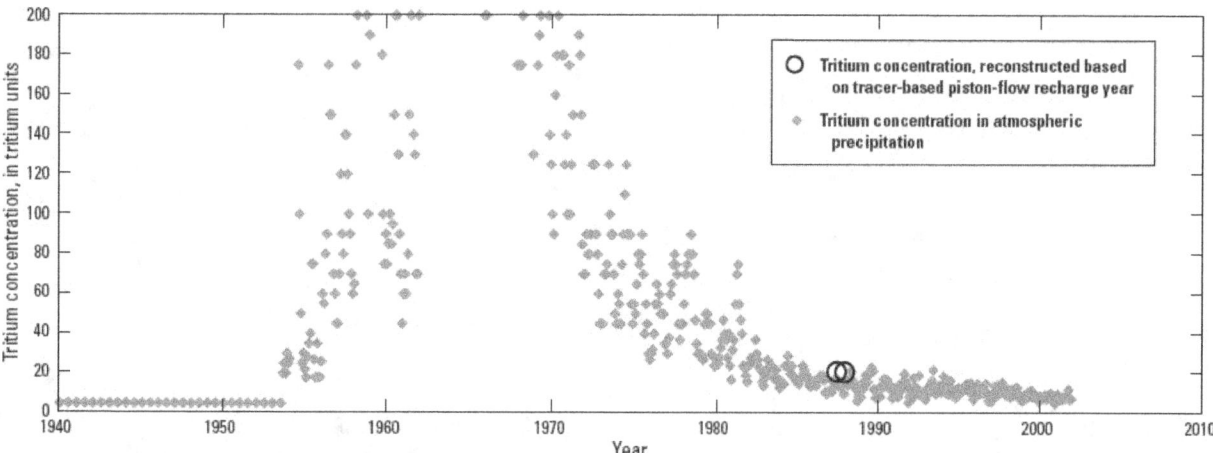

Figure B33. Reconstructed tritium concentrations (using $^3H/^3He$-based ages) and tritium in atmospheric precipitation, CCPTLUSOR1b network, CCYK Study Unit.

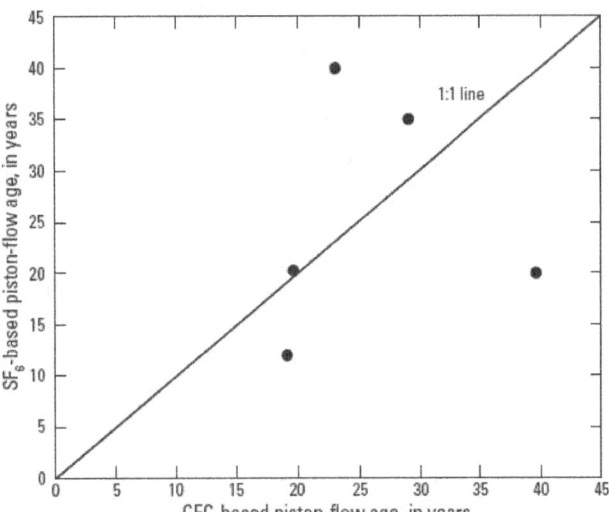

Figure B34 SF$_6$- versus CFC-based age comparison, CCPTLUSOR1b network, CCYK Study Unit.

CCYK CCPTSUS1b

Samples from five sites in the CCYK Study Unit were collected in 2008 for CFCs, SF$_6$, and ^{3}H/^{3}He (networks and, in parentheses, number of sites):

. CCPTSUS1b (5)

The aquifer is composed of the Columbia River Basalt Group basalts; however, one well is finished in the overburden.

Major dissolved-gas data were available for five sites. Of these five sites, all five were oxic.

Age interpretations from tracer concentrations were made assuming that recharge elevation was equal to the elevation of the water table. Estimates of recharge temperature and excess air were based on major dissolved-gas data.

^{3}H/^{3}He ages were calculated for three sites (two of the three sites required a correction for terrigenic helium), while one site was not datable due to fractionation, and the sample from one site was lost due to high pressure or other laboratory issues.

The raw tracer data, major dissolved-gas data, the ancillary chemical and well construction data that were used in the interpretations, and the piston-flow ages are presented in table B9.

. Advantages associated with these samples:

. Multiple tracers (CFCs, SF$_6$, and ^{3}H/^{3}He, as well as major dissolved gases).

. Disadvantages associated with these samples:

. Public supply wells, so generally high pumping stress.

. Relatively large open intervals ranging from 165 to 741 feet so mixing likely.

. Median penetration of center of open interval into water table was 173.13 feet (not sampling close to the water table, potentially mixing).

. Depth to water (can affect tracer transport to water table):

. Median: 180.00 feet

. Mean: 165.9 feet

. Min: 60.00 feet

. Max: 259.00 feet

. Brief analysis:

The CFC-, SF$_6$-, and ^{3}H/^{3}He-based age gradients for these sites are shown in figures B35, B36, and B37. The age gradients do not show any particular structure. Differences in screen length, recharge source/strength, aquifer heterogeneity, pumping stresses, and the position of the well within the flow system may cause some wells to deviate from the general pattern of increasing age with depth.

The reconstructed ^{3}H plots for CFC-, SF$_6$-, and ^{3}H/^{3}He-based ages are shown in figures B38, B39, and B40. The reconstructions show evidence of unmixed, piston-flow transport for some samples, and mixing for other samples. Additional evidence for mixing is seen in the large discrepancy between the tracer ages for two wells for which CFC- and SF$_6$-based ages are more than 30 years younger than the ^{3}H/^{3}He-based ages.

The SF$_6$- versus CFC-based age comparison and the ^{3}H/^{3}He- versus CFC-based age comparison for this network are shown in figures B41 and B42. The age comparisons show similar results to the age-gradient and tritium reconstruction plots shown above. The CFC- and SF$_6$-based ages compare well, but are significantly younger than the ^{3}H/^{3}He-based ages. The low tritium concentrations and high terrigenic helium concentrations limit the reliability of the ^{3}H/^{3}He-based age dating for this network, and the CFC- and SF$_6$-based ages probably are more reliable.

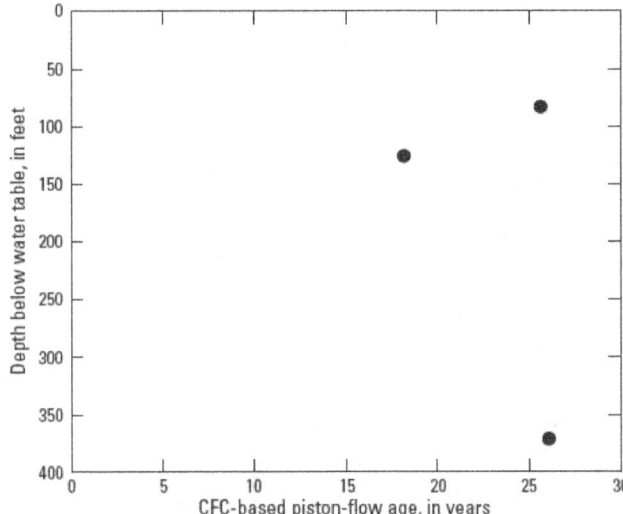

Figure B35. CFC-based age gradient for dated sites from the CCPTSUS1b network, CCYK Study Unit.

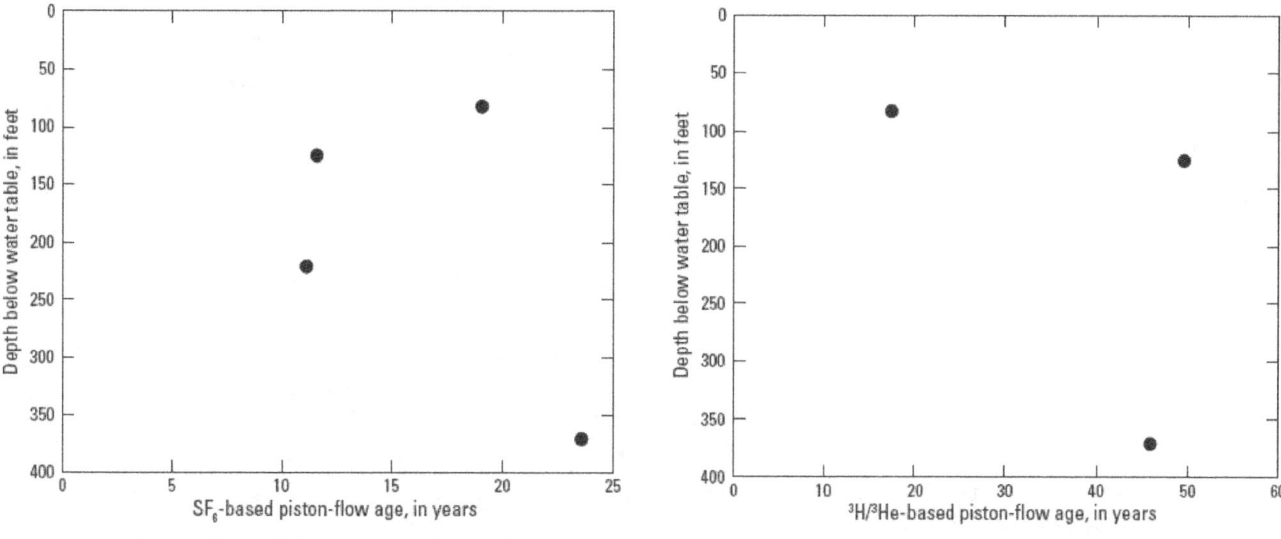

Figure B36. SF$_6$-based age gradient for dated sites from the CCPTSUS1b network, CCYK Study Unit.

Figure B37. ^{3}H/^{3}He-based age gradient for dated sites from the CCPTSUS1b network, CCYK Study Unit.

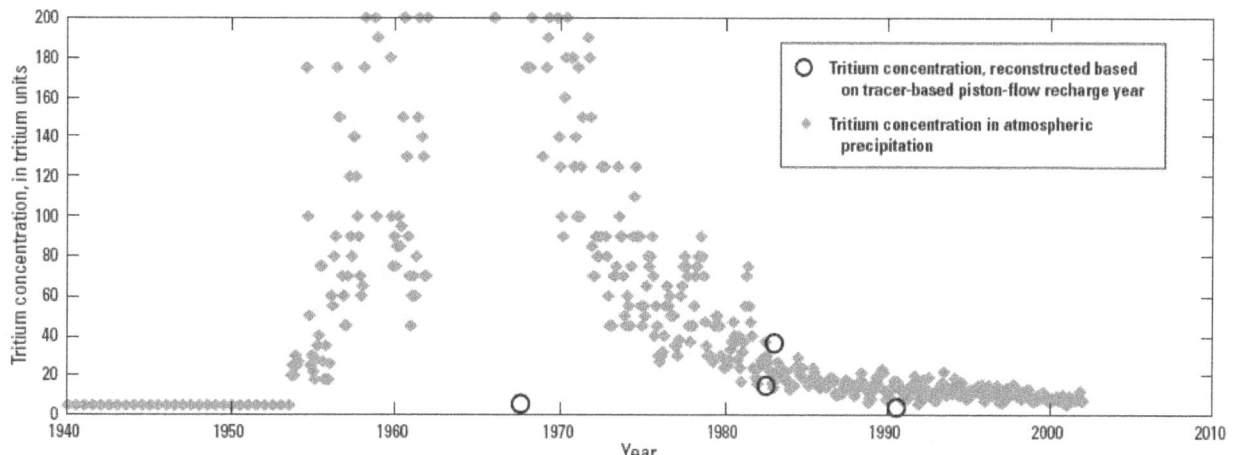

Figure B38. Reconstructed tritium concentrations (using CFC-based ages) and tritium in atmospheric precipitation, CCPTSUS1b network, CCYK Study Unit.

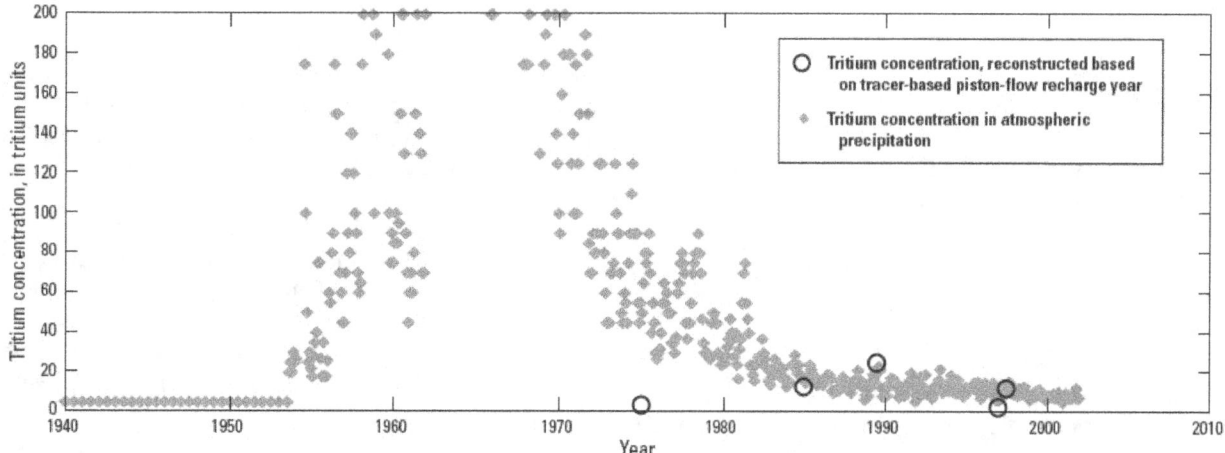

Figure B39. Reconstructed tritium concentrations (using SF$_6$-based ages) and tritium in atmospheric precipitation, CCPTSUS1b network, CCYK Study Unit.

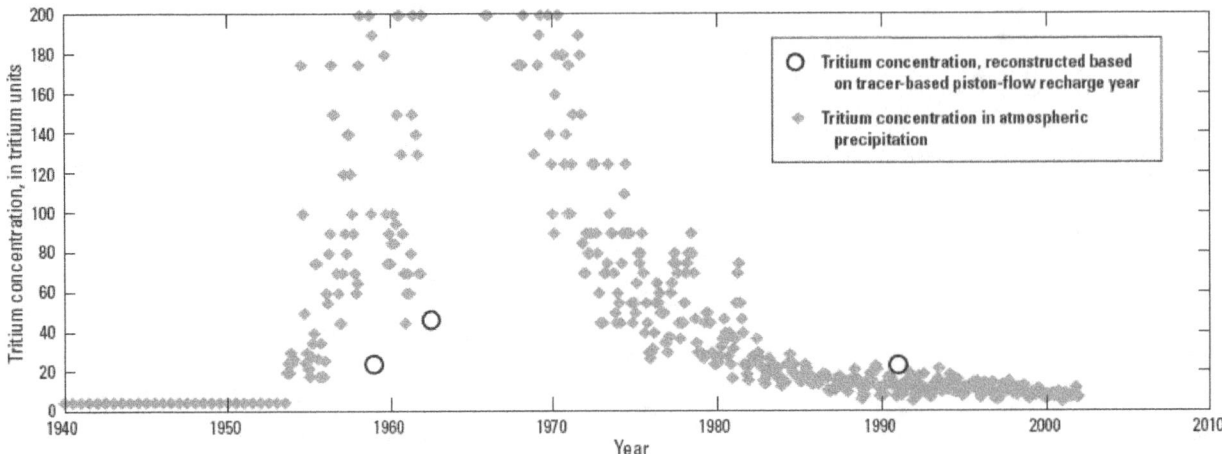

Figure B40.   Reconstructed tritium concentrations (using $^3H/^3He$-based ages) and tritium in atmospheric precipitation, CCPTSUS1b network, CCYK Study Unit.

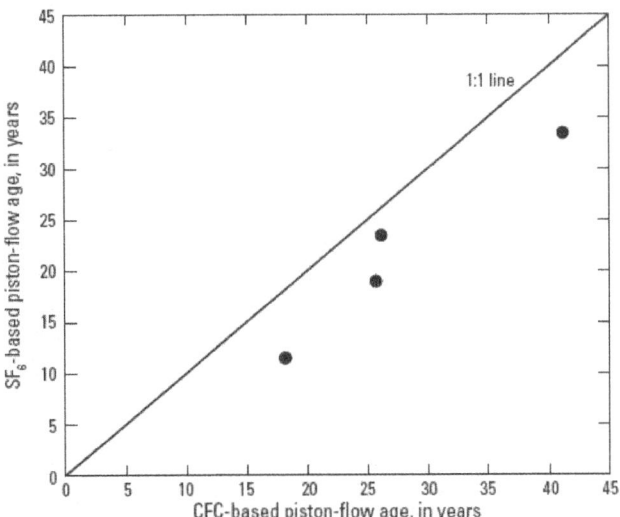

Figure B41. SF_6- versus CFC-based age comparison, CCPTSUS1b network, CCYK Study Unit.

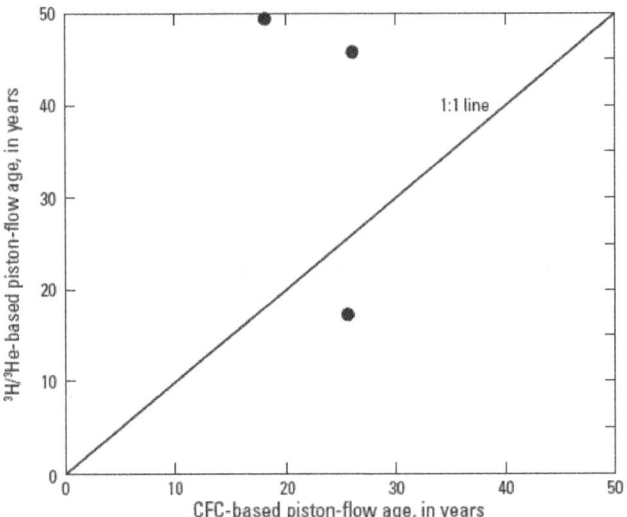

Figure B42. $^3H/^3He$- versus CFC-based age comparison, CCPTSUS1b network, CCYK Study Unit.

CONN SUS2

Samples from five sites in the CONN Study Unit were collected in 2007 for $^3H/^3He$ (networks and, in parentheses, number of sites):

.   SUS2 (5)

The aquifer is composed of sand, silt, and gravel in a glaciated region.

Major dissolved-gas data were available for eight sites. Of these eight sites, six were oxic and two were suboxic.

Age interpretations from tracer concentrations were made assuming that recharge elevation was equal to the elevation of the water table. Estimates of recharge temperature and excess air were based on major dissolved-gas data, with recharge temperature and excess air at suboxic sites being constrained using median excess air at oxic sites.

$^3H/^3He$ ages were calculated for three sites (all three sites required a correction for terrigenic He), while two sites were not datable due to fractionation.

The raw tracer data, major dissolved-gas data, the ancillary chemical and well construction data that were used in the interpretations, and the piston-flow ages are presented in table B10.

Advantages associated with these samples:

- Relatively short open intervals ranging from 1 to 12.8 feet so mixing minimized.

- $^3H/^3He$, as well as major dissolved gases.

Disadvantages associated with these samples:

- Mixture of domestic and monitoring wells, so relatively low pumping stress.

- Median penetration of center of open interval into water table was 49 feet (not sampling close to the water table, potentially mixing).

Depth to water (can affect tracer transport to water table):

- Median: 34.04 feet

- Mean: 43.59 feet

- Min: 5.94 feet

- Max: 100.36 feet

Brief analysis:

The $^3H/^3He$-based age gradient for these sites is shown in figure B43. There is an inverse age gradient, in which tracer-based piston-flow ages are younger at greater depth. These sites are located across multiple states and do not share a common flow path, so the age gradient presented here is not a valid measure of age stratification for this aquifer.

The reconstructed ^3H plot for $^3H/^3He$-based ages is shown in figure B44. There is excellent agreement between the ^3H input function and the reconstructed ages.

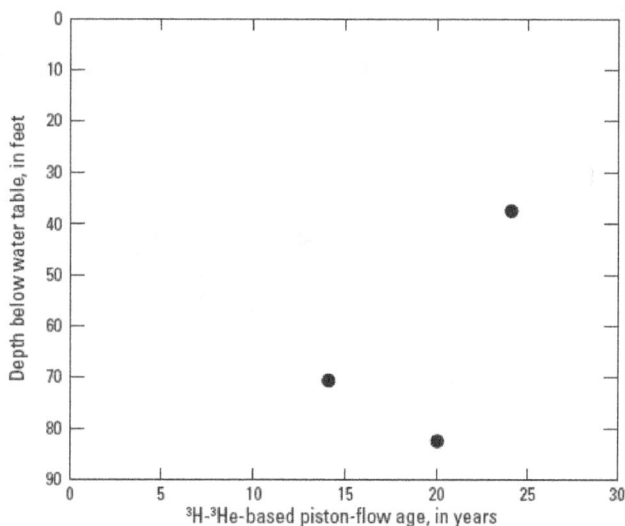

Figure B43. $^3H/^3He$-based age gradient for dated sites from the SUS2 network, CONN Study Unit.

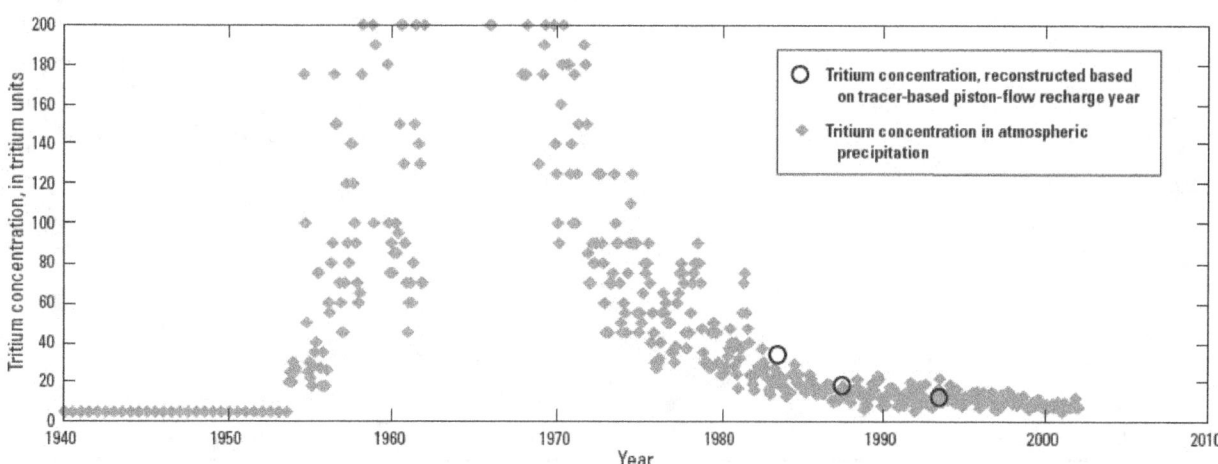

Figure B44. Reconstructed tritium concentrations (using $^3H/^3He$-based ages) and tritium in atmospheric precipitation, SUS2 network, CONN Study Unit.

EIWA FPSAG2

Samples from 9 sites in the EIWA Study Unit were collected in 2007 for CFCs, SF_6, and $^3H/^3He$ (networks and, in parentheses, number of sites):

. FPSAG2 (CFCs, 2; SF_6, 8; $^3H/^3He$, 5).

The aquifer is composed of sand and gravel in a glaciated region.

Major dissolved-gas data were available for 9 sites.

Age interpretations from tracer concentrations were made assuming that recharge elevation was equal to the elevation of the water table, that recharge temperature was equal to mean annual air temperature $+1°C$, and that excess air concentrations were 2 cc STP/kg.

$^3H/^3He$ ages could not be calculated for this network as a result of fractionation.

The raw tracer data, major dissolved-gas data, the ancillary chemical and well construction data that were used in the interpretations, and the piston-flow ages are presented in table B11.

. Advantages associated with these samples:

. Multiple tracers (CFCs, SF_6, and $^3H/^3He$, as well as major dissolved gases).

. Unused wells that no longer exist so low pumping stress.

. Very short open intervals, ranging from 0.31 to 2.31 feet, so mixing minimized.

. Median penetration of center of open interval into water table was 13.5 feet (sampling close to the water table, potentially minimizes mixing).

. Disadvantages associated with these samples:

. Only two CFC samples and they are degraded. All of the $^3H/^3He$ samples are affected by fractionation.

. Depth to water (can affect tracer transport to water table):

. Median: 4.99 feet

. Mean: 5.55 feet

. Min: 1.79 feet

. Max: 8.34 feet

. Brief analysis:

The SF_6-based age gradient for these sites is shown in figure B45. The age gradient has a great deal of scatter. These wells are part of an agricultural network for which irrigation practices may be responsible for the scatter and the relatively old ages at shallow depths.

The reconstructed ^3H plot for SF_6-based ages is shown in figure B46. There is excellent agreement between the ^3H input function and the reconstructed ages.

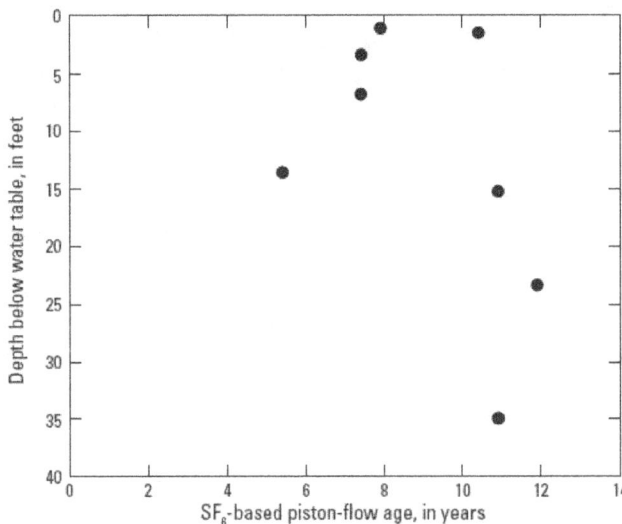

Figure B45. SF_6-based age gradient for dated sites from the FPSAG2 network, EIWA Study Unit.

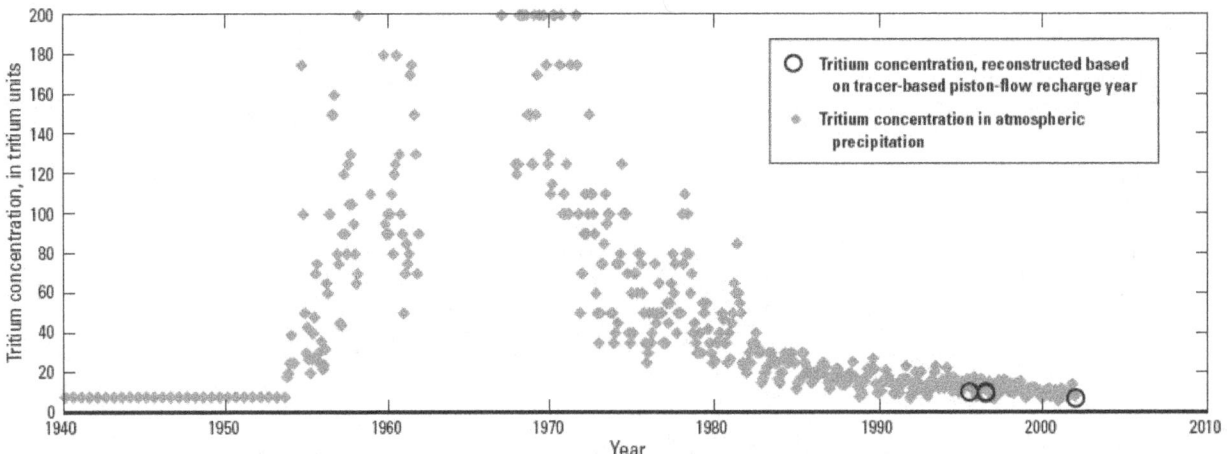

Figure B46.   Reconstructed tritium concentrations (using SF_6-based ages) and tritium in atmospheric precipitation, FPSAG2 network, EIWA Study Unit.

EIWA LUSCR1

Samples from three sites in the EIWA Study Unit were collected in 2007 for CFCs (networks and, in parentheses, number of sites):

. LUSCR1 (3)

The aquifer is composed of sands, gravel, silt, and clay, in a glaciated region.

Major dissolved-gas data were available for all three sites. Of these three sites, one was oxic and two were suboxic.

Age interpretations from tracer concentrations were made assuming that recharge elevation was equal to the elevation of the water table. Estimates of recharge temperature and excess air were based on major dissolved-gas data, with recharge temperature and excess air at suboxic sites being constrained using median excess air at oxic sites from this network as well as other nearby EIWA networks.

The raw tracer data, major dissolved-gas data, the ancillary chemical and well construction data that were used in the interpretations, and the piston-flow ages are presented in table B12.

. Advantages associated with these samples:

. CFCs, as well as major dissolved gases.

. Monitoring wells so generally low pumping stress.

. Short open intervals of 5 feet, so mixing minimized.

. Median penetration of center of open interval into water table was 7.8 feet (sampling close to the water table, potentially minimizing mixing).

. Disadvantages associated with these samples:

. Only 3 samples, and only 1 was datable.

. Depth to water (can affect tracer transport to water table):

. Median: 3.90 feet

. Mean: 3.88 feet

. Min: 2.20 feet

. Max: 5.55 feet

. Brief analysis:

. With only one datable site, no age gradient or ^{3}H reconstruction was attempted.

EIWA SUS2

Samples from 23 sites in the EIWA Study Unit were collected in 2007 for CFCs (networks and, in parentheses, number of sites):

. SUS2 (CFCs, 23).

The aquifer is composed of glacial sands, gravel, and clay.

Major dissolved-gas data were available for 22 sites. Of these 22 sites, 7 were oxic and 15 were suboxic.

Age interpretations from tracer concentrations were made assuming that recharge elevation was equal to the elevation of the water table. Estimates of recharge temperature and excess air were based on major dissolved-gas data, with recharge temperature and excess air at suboxic sites being constrained using median excess air at oxic sites.

The raw tracer data, major dissolved-gas data, the ancillary chemical and well construction data that were used in the interpretations, and the piston-flow ages are presented in table B13.

. Advantages associated with these samples:

. CFCs, as well as major dissolved gases.

. Domestic wells so generally low pumping stress.

. Disadvantages associated with these samples:

. Relatively large open intervals ranging from 4 to 177 feet, with many unknown, so mixing likely.

. Median penetration of center of open interval into water table was 48.69 feet (not sampling close to the water table, potentially mixing).

. Depth to water (can affect tracer transport to water table):

. Median: 10.60 feet

. Mean: 19.64 feet

. Min: 2.10 feet

. Max: 80.70 feet

. Brief analysis:

. In the EIWA network, in areas where clay and shale materials are present, it is common to find water recharged within the last 40 years that has not reached the deeper parts of the alluvial and bedrock aquifers. This could explain why there is an inverted age gradient for the SUS2 samples as shown in figure B47.

The reconstructed ^{3}H plot for CFC-based ages is shown in figure B48. There is good agreement between the ^{3}H input function and the reconstructed ages indicating that the samples represent relatively unmixed, piston-flow transport, with some samples showing effects of dispersion.

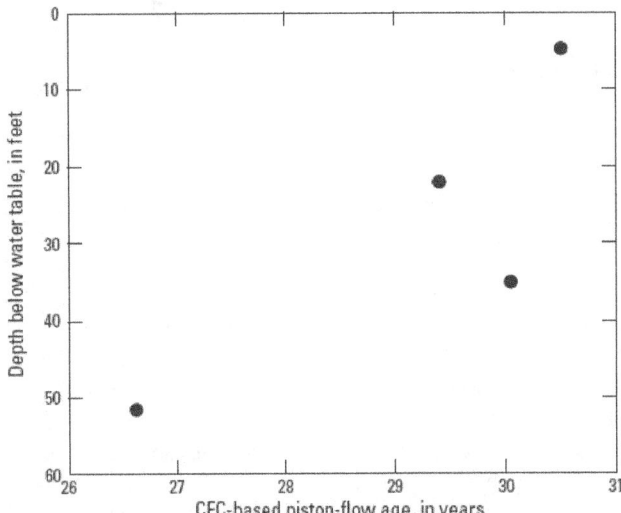

Figure B47.   CFC-based age gradient for dated sites from the SUS2 network, EIWA Study Unit.

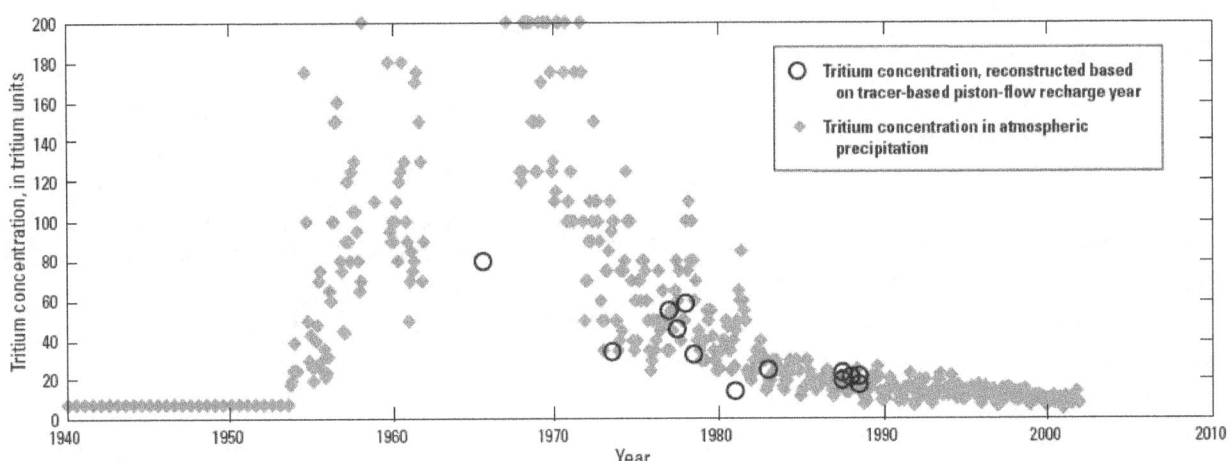

Figure B48.   Reconstructed tritium concentrations (using CFC-based ages) and tritium in atmospheric precipitation, SUS2 network, EIWA Study Unit.

GRSL SUS1a, SUS1b, and LUSRC1

Samples from 10 sites in the GRSL Study Unit were collected in 2006 for CFCs, SF_6, and $^3H/^3He$ (networks and, in parentheses, number of sites):

.   SUS1a (1)

.   SUS1b (4)

.   LUSRC1 (5)

The aquifer is composed of valley-fill deposits of sand and gravel.

Major dissolved-gas data were available for five sites. All five sites were oxic.

Age interpretations from tracer concentrations were made assuming that recharge elevation was equal to the elevation of the water table. Estimates of recharge temperature and excess air were based on major dissolved-gas data. For the sites without major dissolved-gas data, age interpretations from tracer concentrations were made assuming that recharge temperature was equal to mean annual air temperature +1°C, and that excess air concentrations were 2 cc STP/kg because the samples with major dissolved-gas data, while consistent among duplicates, were quite variable from one site to the next.

^{3}H/3He ages were calculated for eight sites (four of the eight sites required a correction for terrigenic helium), while samples from two sites were lost due to high pressure or other laboratory issues.

The raw tracer data, major dissolved-gas data, the ancillary chemical and well construction data that were used in the interpretations, and the piston-flow ages are presented in table B14.

. Advantages associated with these samples:

. Multiple tracers (CFCs, SF^6, and ^{3}H/3He, as well as major dissolved gases).

. Disadvantages associated with these samples:

. Mixture of domestic, unused, irrigation, and commercial wells, so variable pumping rates.

. Relatively large open intervals ranging from 10 to 94 feet so mixing likely.

. Median penetration of center of open interval into water table was 56.74 feet (not sampling close to the water table, potentially mixing).

. Depth to water (can affect tracer transport to water table):

. Median: 70.22 feet

. Mean: 75.91 feet

. Min: 21.60 feet

. Max: 165.40 feet

. Brief analysis:

. The CFC-, SF$_6$-, and ^{3}H/3He-based age gradients for these sites are shown in figures B49, B50, and B51. The age gradients show no clear trend, however, the ^{3}H/3He ages are somewhat younger in the shallow wells indicating possible helium loss near the water table. Differences in screen length, recharge source/strength, aquifer heterogeneity, pumping stresses, and the position of the well within the flow system may cause some wells to deviate from the general pattern of increasing age with depth.

The reconstructed ^{3}H plots for CFC-, SF$_6$-, and ^{3}H/3He-based ages are shown in figures B52, B53, and B54. The reconstructions show evidence of unmixed, piston-flow transport for some samples, and mixing for other samples, as would be expected for such a wide variety of well types and depths.

The ^{3}H/3He- versus SF$_6$-based age comparison for this network is shown in figure B55. The age comparison is limited by the fact that there are only three samples and only one same shows age agreement.

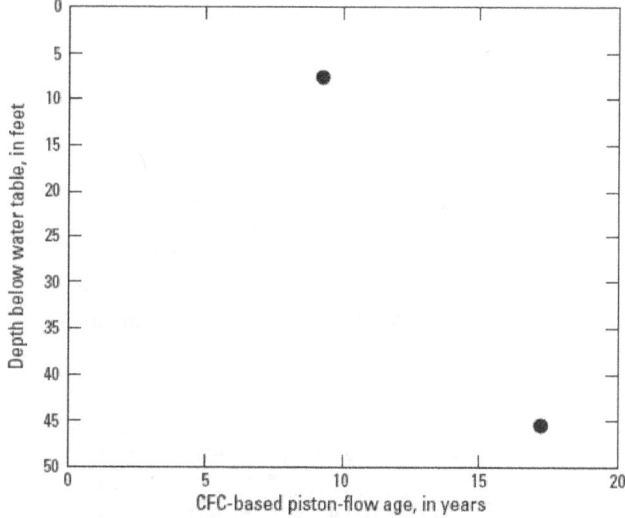

Figure B49. CFC-based age gradient for dated sites from the SUS1a, SUS1b, and LUSRC1 networks, GRSL Study Unit.

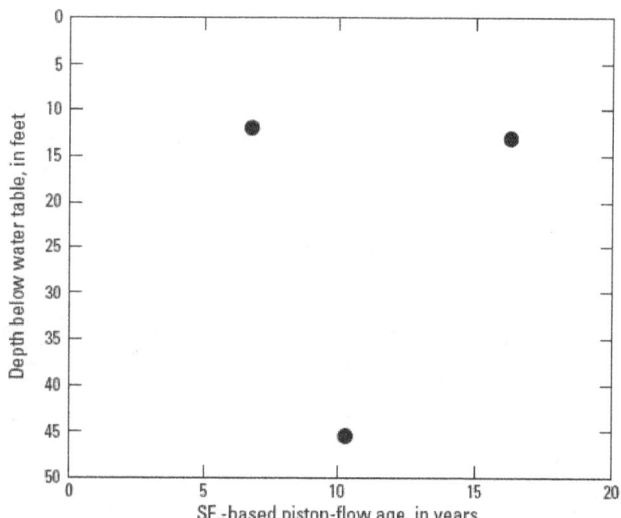

Figure B50. SF$_6$-based age gradient for dated sites from the SUS1a, SUS1b, and LUSRC1 networks, GRSL Study Unit.

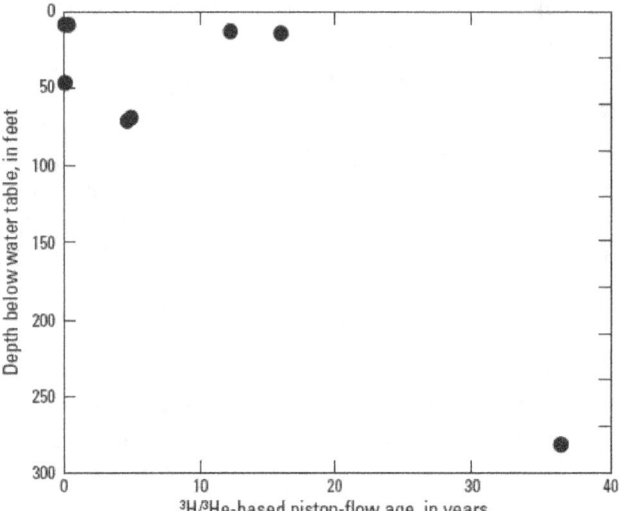

Figure B51. ^{3}H/^{3}He-based age gradient for dated sites from the SUS1a, SUS1b, and LUSRC1 networks, GRSL Study Unit.

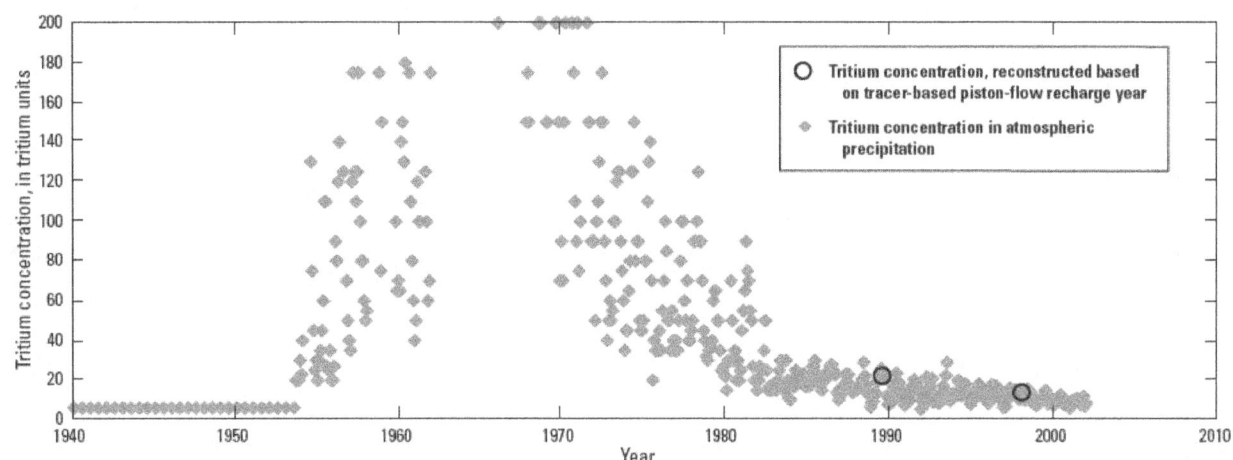

Figure B52.   Reconstructed tritium concentrations (using CFC-based ages) and tritium in atmospheric precipitation, SUS1a, SUS1b, and LUSRC1 networks, GRSL Study Unit.

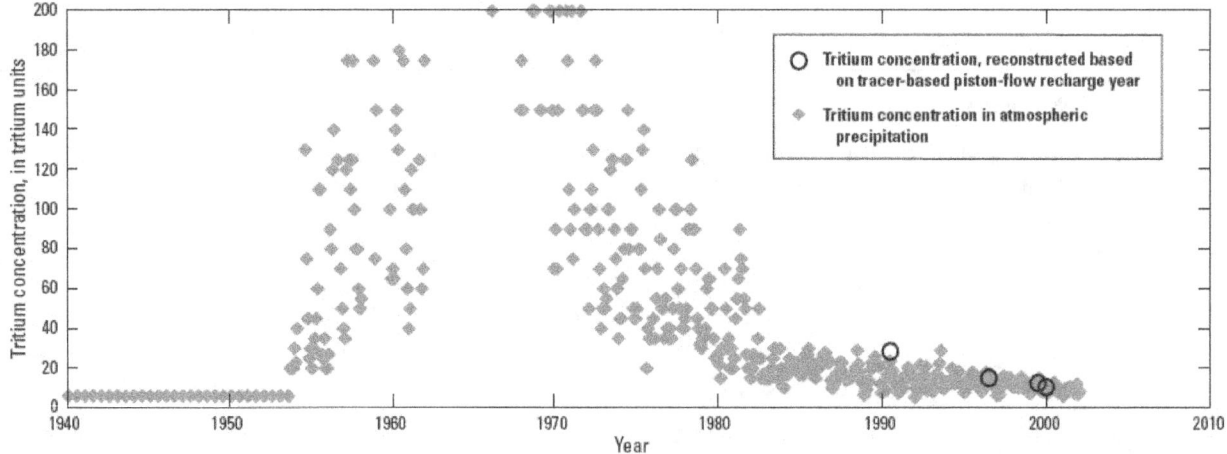

Figure B53.   Reconstructed tritium concentrations (using SF$_6$-based ages) and tritium in atmospheric precipitation, SUS1a, SUS1b, and LUSRC1 networks, GRSL Study Unit.

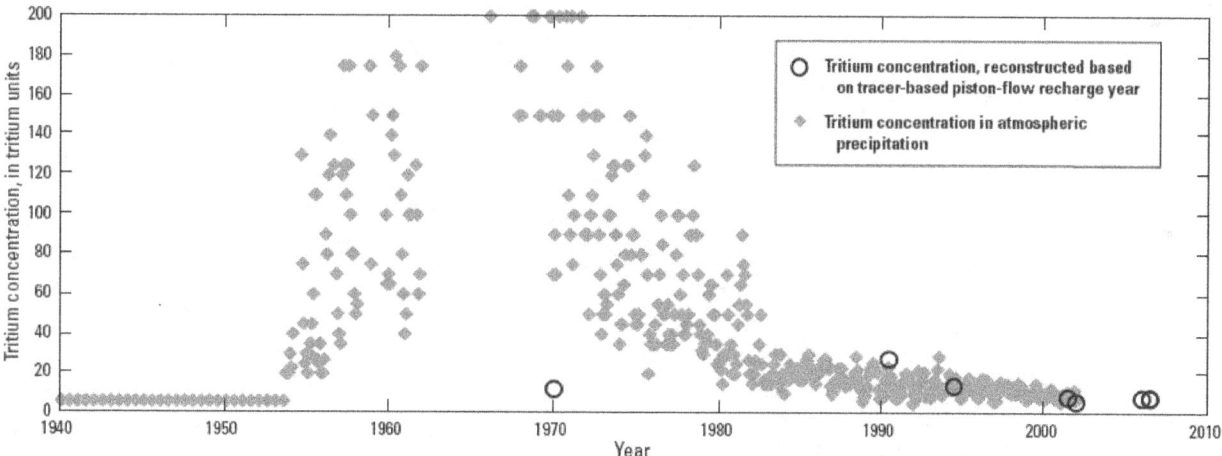

Figure B54. Reconstructed tritium concentrations (using ^{3}H/^{3}He-based ages) and tritium in atmospheric precipitation, SUS1a, SUS1b, and LUSRC1 networks, GRSL Study Unit.

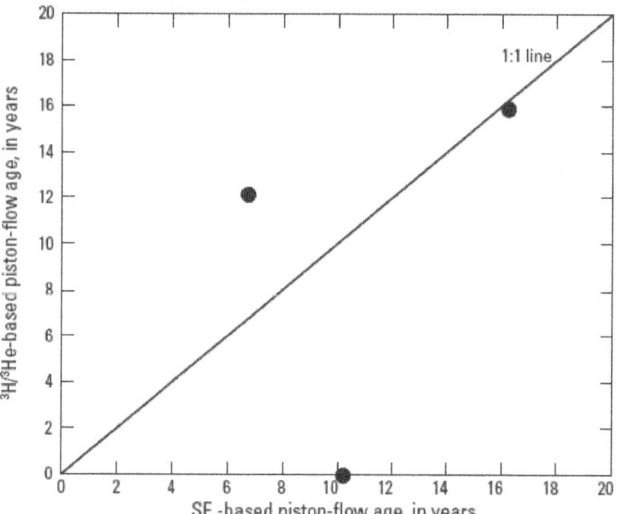

Figure B55. ^{3}H/^{3}He- versus SF$_6$-based age comparison, SUS1a, SUS1b, and LUSRC1 networks, GRSL Study Unit.

GRSL SUS1a, SUS1b, SUS2, and SUS3

Samples from 36 sites in the GRSL Study Unit were collected in 2008 for CFCs, SF$_6$, and ^{3}H/^{3}He (networks and, in parentheses, number of sites):

.   SUS1a (7)

.   SUS1b (7)

.   SUS2 (8)

.   SUS3 (14)

The aquifer is composed of Basin and Range basin-fill sands, gravel, and clay.

Major dissolved-gas data were available for all 36 sites. Of these 36 sites, 30 were oxic and 6 were suboxic.

Age interpretations from tracer concentrations were made assuming that recharge elevation was equal to the elevation of the water table. Estimates of recharge temperature and excess air were based on major dissolved-gas data, with recharge temperature and excess air at suboxic sites being constrained using median excess air at oxic sites.

^{3}H/^{3}He ages were calculated for 20 sites (9 of the 20 sites required a correction for terrigenic helium), while 6 site were not datable because tritium concentrations were too low. 2 sites were not datable because of fractionation, and samples from 5 sites were lost due to high pressure or other laboratory issues. Site S10 had a tritium concentration from a sample taken in 1998, which was decayed to 2008 for use in the ^{3}H/^{3}He spreadsheet.

The raw tracer data, major dissolved-gas data, the ancillary chemical and well construction data that were used in the interpretations, and the piston-flow ages are presented in table B15.

.   Advantages associated with these samples:

.   Multiple tracers (CFCs, SF$_6$, and ^{3}H/^{3}He, as well as major dissolved gases).

.   Disadvantages associated with these samples:

.   Mixture of domestic, irrigation, commercial, and public supply wells, so variable pumping rates.

.   Relatively large open intervals ranging from 5 to 733 feet so mixing likely.

.   Median penetration of center of open interval into water table was 170.81 feet (not sampling close to the water table, potentially mixing).

. Depth to water (can affect tracer transport to water table):

    Median: 124.75 feet

    Mean: 143.67 feet

    Min: 10.83 feet

    Max: 430.00 feet

. Brief analysis:

    The CFC-, SF_6-, and $^3H/^3He$-based age gradients for these sites are shown in figures B56, B57, and B58. The age gradients do not show any particular structure. Differences in screen length, recharge source/strength, aquifer heterogeneity, pumping stresses, and the position of the well within the flow system may cause some wells to deviate from the general pattern of increasing age with depth.

The reconstructed ^3H plots for SF_6- and $^3H/^3He$-based ages are shown in figures B59 and B60. The reconstructions show that most samples are of mixed-age water as would be expected for such a wide variety of well types and depths. The $^3H/^3He$-based reconstruction, however, shows evidence of capturing the period of time around the bomb peak, with numerous samples with ages in the 1950 and 1960s exhibiting elevated ^3H reconstructions.

The $^3H/^3He$- versus SF_6-based age comparison for this network is shown in figure B61. The age comparison shows significant disagreement among tracers suggesting groundwater is generally of mixed age.

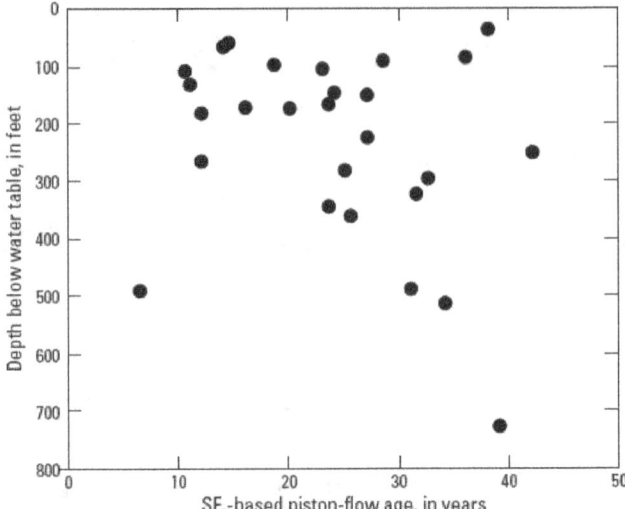

Figure B57. SF_6-based age gradient for dated sites from the SUS1a, SUS1b, SUS2, and SUS3 networks, GRSL Study Unit.

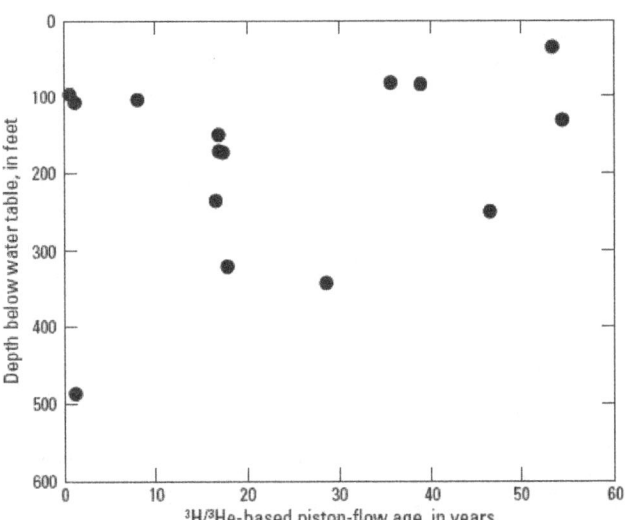

Figure B58. $^3H/^3He$-based age gradient for dated sites from the SUS1a, SUS1b, SUS2, and SUS3 networks, GRSL Study Unit.

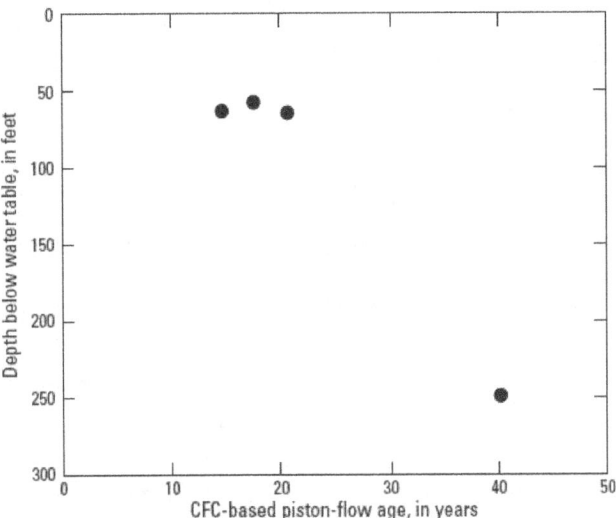

Figure B56.   CFC-based age gradient for dated sites from the SUS1a, SUS1b, SUS2, and SUS3 networks, GRSL Study Unit.

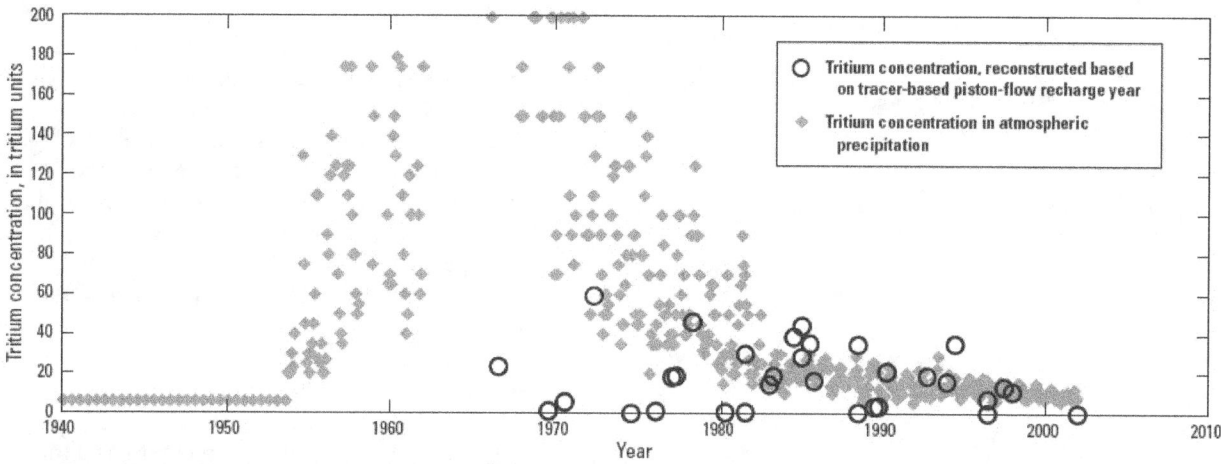

Figure B59. Reconstructed tritium concentrations (using SF_6-based ages) and tritium in atmospheric precipitation, SUS1a, SUS1b, SUS2, and SUS3 networks, GRSL Study Unit.

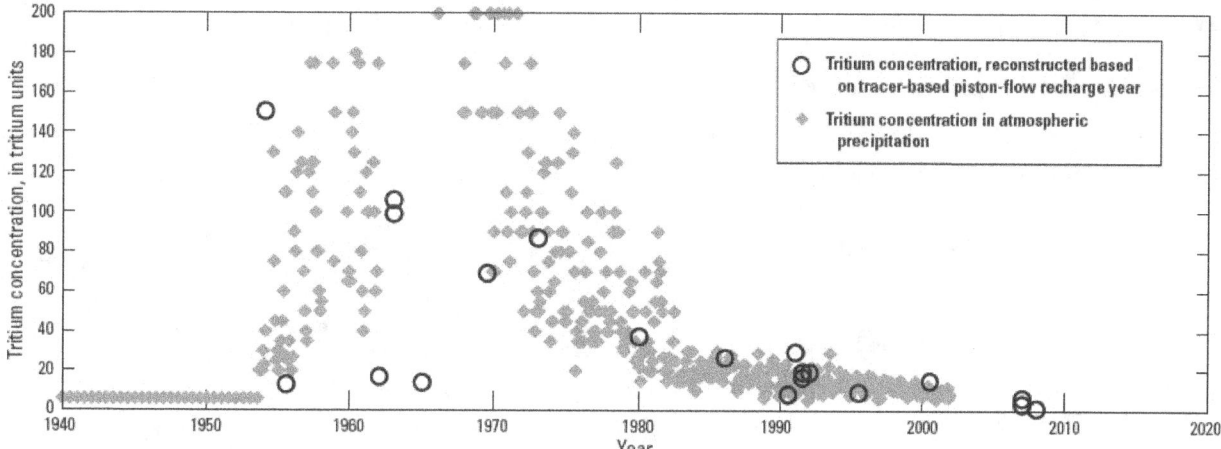

Figure B60. Reconstructed tritium concentrations (using $^3H/^3He$-based ages) and tritium in atmospheric precipitation, SUS1a, SUS1b, SUS2, and SUS3 networks, GRSL Study Unit.

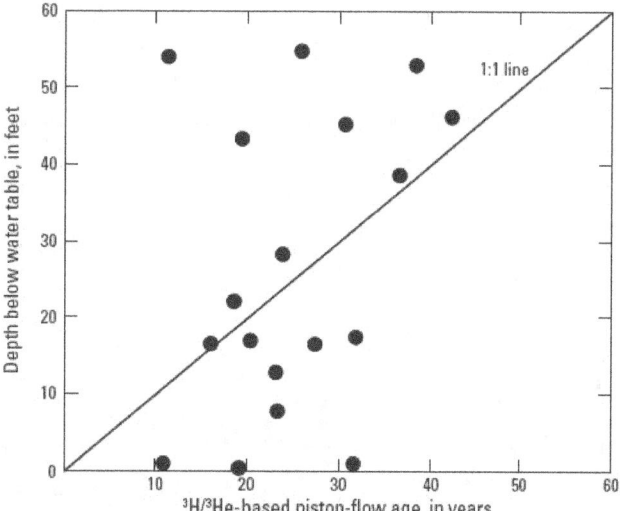

Figure B61. $^3H/^3He$- versus SF_6-based age comparison, SUS1a, SUS1b, SUS2, and SUS3 networks, GRSL Study Unit.

LERI FPSRC1

Samples from 19 sites in the LERI Study Unit were collected in 2008 for SF_6 (networks and, in parentheses, number of sites):

. FPSRC1 (19)

The aquifer is composed of outwash sand, silt, and gravel, with one well completed in the Mississippian Shale.

Major dissolved-gas data were available for 20 sites. Of these 20 sites, 7 were oxic and 13 were suboxic.

Age interpretations from tracer concentrations were made assuming that recharge elevation was equal to the elevation of the water table. Estimates of recharge temperature and excess air were based on major dissolved-gas data, with recharge temperature and excess air at suboxic sites being constrained using median excess air at oxic sites.

The raw tracer data, major dissolved-gas data, the ancillary chemical and well construction data that were used in the interpretations, and the piston-flow ages are presented in table B16.

. Advantages associated with these samples:

  . SF_6, as well as major dissolved gases.

  . Relatively short open intervals ranging from 2 to 5 feet so mixing likely minimized.

  . Median penetration of center of open interval into water table was 16.47 feet (sampling close to the water table, potentially minimizing mixing).

. Disadvantages associated with these samples:

  . Mixture of domestic, public supply, and monitoring wells, so variable pumping rates.

. Depth to water (can affect tracer transport to water table):

  . Median: 1.88 feet

  . Mean: 5.67 feet

  . Min: 0.25 feet

  . Max: 5.00 feet

. Brief analysis:

The SF_6-based age gradient for these sites is shown in figure B62. The age gradient shows a general trend of increasing age with depth. Differences in screen length, recharge source/strength, aquifer heterogeneity, pumping stresses, and the position of the well within the flow system may cause some wells to deviate from the general pattern of increasing age with depth. For example, the sample with the young age for the deepest well is from a domestic well finished in a different aquifer than the rest of the wells.

The reconstructed ^3H plot for SF_6-based ages is shown in figure B63. The reconstruction shows evidence of unmixed, piston-flow transport for some samples, and mixing for other samples. Three of the four wells with reconstructed tritium values that plot well below the input curve are domestic or public supply wells with longer open intervals than the remaining unused monitoring wells. One of these wells, Well 4-205 is the well finished in shale.

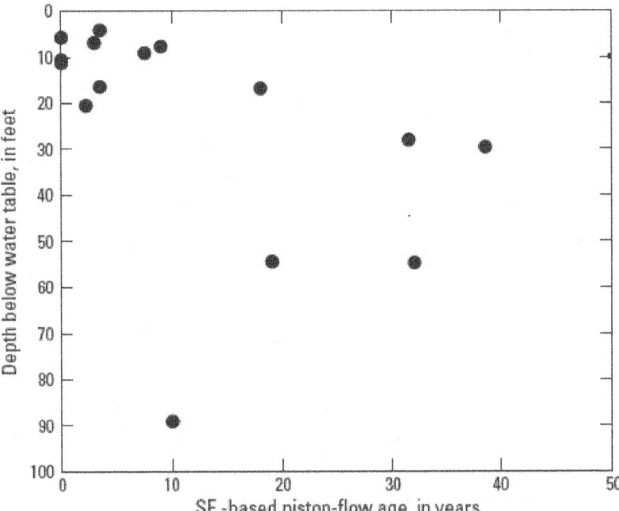

Figure B62. SF_6-based age gradient for dated sites from the FPSRC1 network, LERI Study Unit.

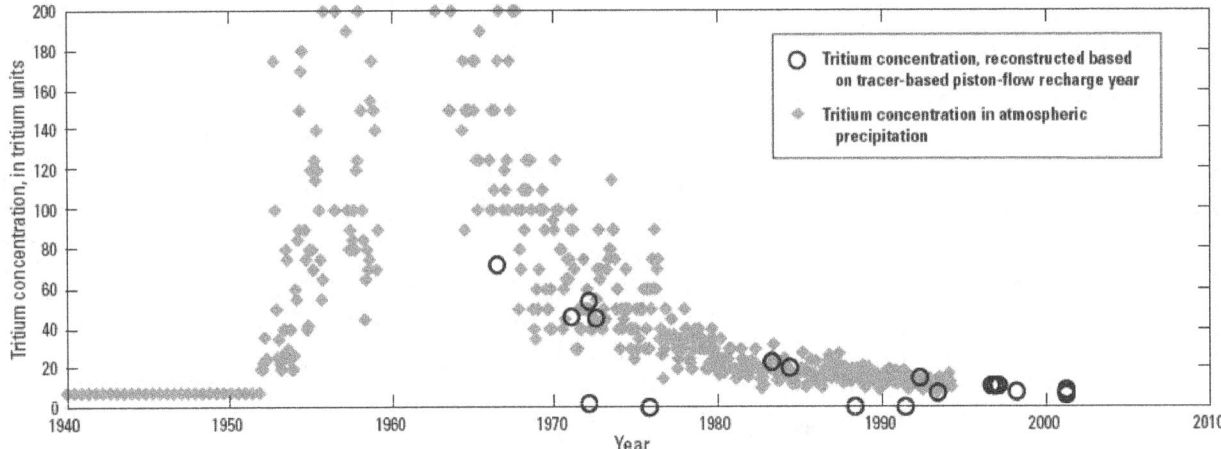

Figure B63. Reconstructed tritium concentrations (using SF_6-based ages) and tritium in atmospheric precipitation, FPSRC1 network, LERI Study Unit.

LERI LUSRC1 and REFOT1

Samples from 25 sites in the LERI Study Unit were collected in 2007 for ^{3}H/^{3}He (networks and, in parentheses, number of sites):

. LUSRC1 (23)

. REFOT1 (2)

The aquifer is composed of outwash sands, silts, clays, and gravel.

Major dissolved-gas data were available for five sites. Of these five sites, four were oxic and one was suboxic.

Age interpretations from tracer concentrations were made assuming that recharge elevation was equal to the elevation of the water table. Estimates of recharge temperature and excess air were based on major dissolved-gas data, with recharge temperature and excess air at suboxic sites being constrained using median excess air at oxic sites. For the one suboxic sample, no correction for denitrification was used because the recharge temperature would have been unreasonably low.

^{3}H/^{3}He ages were calculated for 17 sites (4 of the 17 sites required a correction for terrigenic helium), while 6 sites were not datable because of fractionation, and samples from 2 sites were lost due to high pressure.

The raw tracer data, major dissolved-gas data, the ancillary chemical and well construction data that were used in the interpretations, and the piston-flow ages are presented in table B17.

. Advantages associated with these samples:

. ^{3}H/^{3}He, as well as major dissolved gases.

. Monitoring wells, so generally low pumping stress.

. Relatively short open intervals ranging from 4.5 to 8.1 feet so mixing likely minimized.

. Median penetration of center of open interval into water table was 9.21 feet (sampling close to the water table, potentially minimizing mixing).

. Disadvantages associated with these samples:

. None.

. Depth to water (can affect tracer transport to water table):

. Median: 12.27 feet

. Mean: 16.07 feet

. Min: 4.95 feet

. Max: 58.65 feet

. Brief analysis:

. The ^{3}H/^{3}He-based age gradient for these sites is shown in figure B64. The age gradient shows a general increase in age with depth. Differences in screen length, recharge source/strength, aquifer heterogeneity, pumping stresses, and the position of the well within the flow system may cause some wells to deviate from the general pattern of increasing age with depth. The zero age does not intercept at zero depth likely as a result of helium loss in the unsaturated zone, which ranges in depth from 5 to 60 feet below land surface.

The reconstructed ^{3}H plot for ^{3}H/^{3}He-based ages is shown in figure B65. The reconstruction shows evidence of unmixed, piston-flow transport, which is consistent with the age gradient for this network.

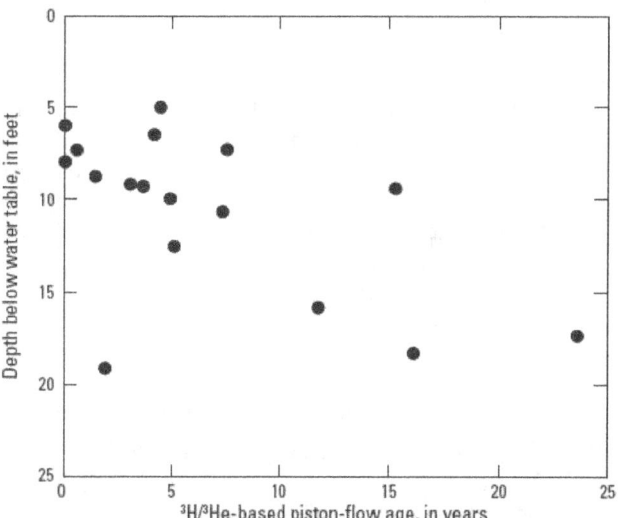

Figure B64. ^{3}H/^{3}He-based age gradient for dated sites from the LUSRC1 and REFOT1 networks, LERI Study Unit.

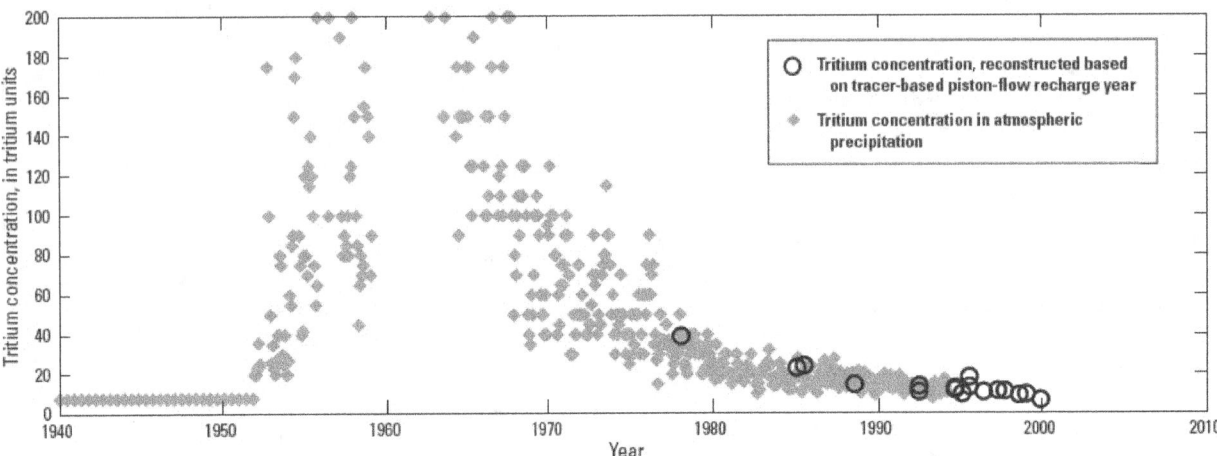

Figure B65. Reconstructed tritium concentrations (using $^3H/^3He$-based ages) and tritium in atmospheric precipitation, LUSRC1 and REFOT1 networks, LERI Study Unit.

LERI SUS1

Samples from 16 sites in the LERI Study Unit were collected in 2007 for CFCs (networks and, in parentheses, number of sites):

- SUS1 (16)

The aquifer is composed of sand and gravel.

Major dissolved-gas data were available for all 16 sites. Of these 16 sites, all 16 were suboxic and samples from three sites were degassed.

Age interpretations from tracer concentrations were made assuming that recharge elevation was equal to the elevation of the water table, that recharge temperature was equal to the mean annual air temperature +1°C, and that excess air concentrations were 2 cc STP/kg.

The raw tracer data, major dissolved-gas data, the ancillary chemical and well construction data that were used in the interpretations, and the piston-flow ages are presented in table B18.

- Advantages associated with these samples:

  - CFCs, as well as major dissolved gases.

  - Mixture of domestic and monitoring wells, so generally low pumping stress.

  - Relatively short open intervals ranging from 3 to 10 feet so mixing likely minimized.

- Disadvantages associated with these samples:

  - Median penetration of center of open interval into water table was 52.49 feet (not sampling close to the water table, potentially mixing).

  - The CFCs were degraded and no age-dating was possible for these sites because only CFC samples were taken.

- Depth to water (can affect tracer transport to water table):

  - Median: 22.79 feet

  - Mean: 26.84 feet

  - Min: 1.87 feet

  - Max: 69.65 feet

- Brief analysis:

  - No tracer interpretations could be made for this network as a result of the suboxic conditions.

LINJ LUSRC1, SUS2, REFFO1, and FPSOT3

Samples from 12 sites in the LINJ Study Unit were collected in 2007 for SF_6 (networks and, in parentheses, number of sites):

- LUSRC1 (4)

- SUS2 (5)

- REFFO1 (1)

- FPSOT3 (2)

The aquifer is composed of sand, clay, and gravel of the Cohansey Sand-Kirkwood Formation.

Age interpretations from tracer concentrations were made assuming that recharge elevation was equal to the elevation of the water table, that recharge temperature was equal to the mean annual air temperature +1°C, and that excess air concentrations were 2 cc STP/kg.

The raw tracer data, the ancillary chemical and well construction data that were used in the interpretations, and the piston-flow ages are presented in table B19.

- Advantages associated with these samples:

  - None.

. Disadvantages associated with these samples:

. No multiple tracers or major dissolved gases.

. Mixture of domestic, monitoring, commercial, and public supply wells, so variable pumping rates.

. Relatively large open intervals ranging from 2 to 34.25 feet so mixing likely.

. Median penetration of center of open interval into water table was 61.61 feet (not sampling close to the water table, potentially mixing).

. No tritium analyses.

. Depth to water (can affect tracer transport to water table):

. Median: 14.72 feet

. Mean: 25.90 feet

. Min: 8.85 feet

. Max: 51.39 feet

. Brief analysis:

The SF_6-based age gradient for these sites is shown in figure B66. The age gradient shows a general trend of increasing age with depth. Differences in screen length, recharge source/strength, aquifer heterogeneity, pumping stresses, and the position of the well within the flow system may cause some wells to deviate from the general pattern of increasing age with depth.

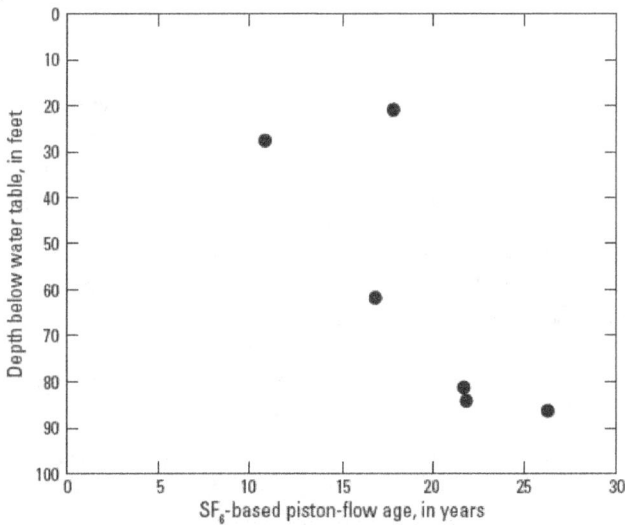

Figure B66. SF_6-based age gradient for dated sites from the LUSRC1, SUS2, REFFO1, and FPSOT3 networks, LINJ Study Unit.

LINJ LUSRC2

Samples from 27 sites in the LINJ Study Unit were collected in 2006 for SF_6 and $^3H/^3He$ (networks and, in parentheses, number of sites):

. LUSRC2 (SF_6, 26; $^3H/^3He$, 8)

The aquifer is composed of undifferentiated glacial deposits.

Major dissolved-gas data were available for nine sites. Of these nine sites, five were oxic and four were suboxic.

Age interpretations from tracer concentrations were made assuming that recharge elevation was equal to the elevation of the water table. Estimates of recharge temperature and excess air were based on major dissolved-gas data, with recharge temperature and excess air at suboxic sites being constrained using median excess air at oxic sites.

$^3H/^3He$ ages were calculated for four sites (three of the four sites required a correction for terrigenic helium), while one site was not datable because of fractionation, and samples from three sites were lost due to high pressure or other laboratory issues.

The raw tracer data, major dissolved-gas data, the ancillary chemical and well construction data that were used in the interpretations, and the piston-flow ages are presented in table B20.

. Advantages associated with these samples:

. Multiple tracers (SF_6 and $^3H/^3He$, as well as major dissolved gases).

. Monitoring wells, so generally low pumping stress.

. Relatively short open intervals ranging from 3 to 10 feet so mixing likely minimized.

. Median penetration of center of open interval into water table was 19.57 feet (sampling close to the water table, potentially minimizing mixing).

. Disadvantages associated with these samples:

. No tritium analyses.

. Depth to water (can affect tracer transport to water table):

. Median: 25.61 feet

. Mean: 30.94 feet

. Min: 4.09 feet

. Max: 84.50 feet

. Brief analysis:

. The SF_6- and $^3H/^3He$-based age gradients for these sites are shown in figure B67 and B68. Both age gradients show a general increase in age with depth. Differences in screen length, recharge source/strength, aquifer heterogeneity, pumping stresses, and the position of the well within the flow system may cause some wells to deviate from the general pattern of increasing age with depth.

The $^3H/^3He$- versus SF_6-based age comparison for this network is shown in figure B69. The age comparison is limited by the low number of samples, but is at least consistent for two of the three samples.

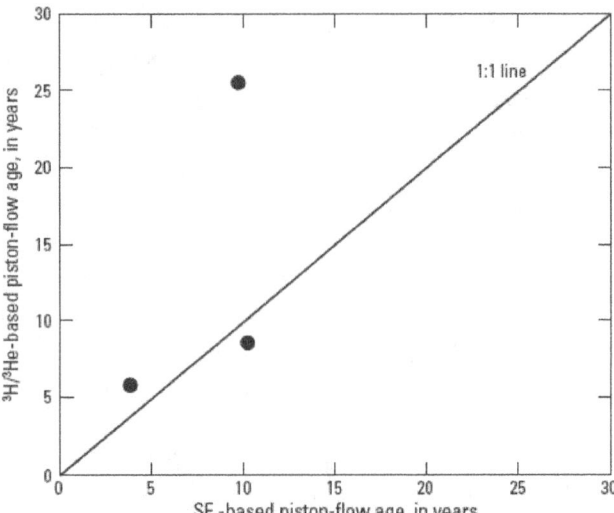

Figure B69. $^3H/^3He$- versus SF_6-based age comparison, LUSRC2 network, LINJ Study Unit.

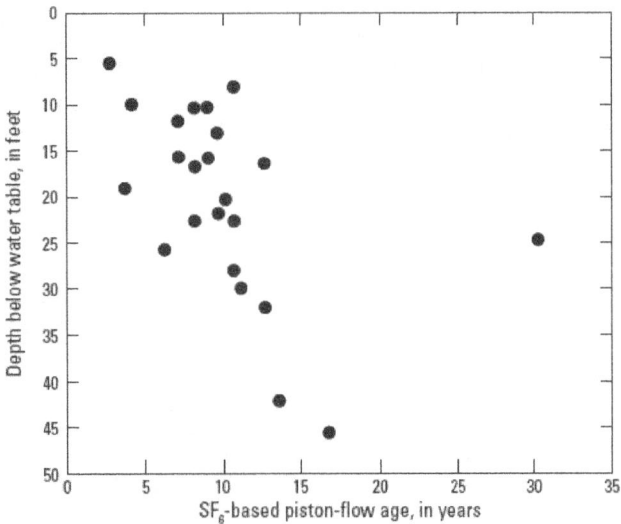

Figure B67. SF_6-based age gradient for dated sites from the LUSRC2 network, LINJ Study Unit.

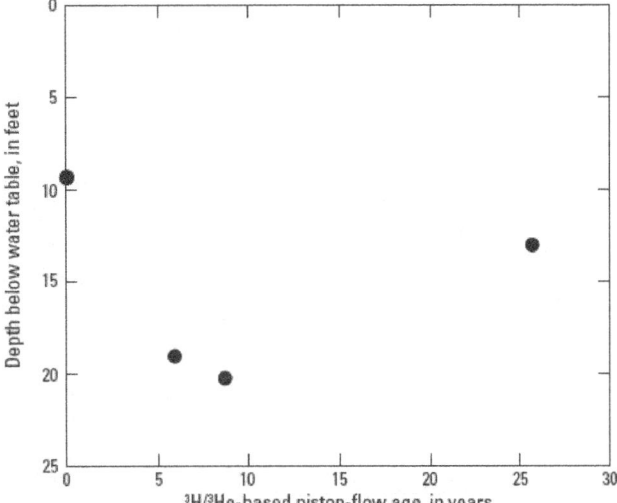

Figure B68. $^3H/^3He$-based age gradient for dated sites from the LUSRC2 network, LINJ Study Unit.

LINJ SUS2

Samples from 11 sites in the LINJ Study Unit were collected in 2006 for SF_6 (networks and, in parentheses, number of sites):

. SUS2 (11)

The aquifer is composed of Basin and Range basin-fill sand, gravel, clay, and silt of the Cohansey Sand-Kirkwood Formation.

Major dissolved-gas data were available for nine sites. Of these nine sites, seven were oxic and two were suboxic.

Age interpretations from tracer concentrations were made assuming that recharge elevation was equal to the elevation of the water table. Estimates of recharge temperature and excess air were based on major dissolved-gas data, with recharge temperature and excess air at suboxic sites being constrained using median excess air at oxic sites.

The raw tracer data, major dissolved-gas data, the ancillary chemical and well construction data that were used in the interpretations, and the piston-flow ages are presented in table B21.

. Advantages associated with these samples:

. SF_6, as well as major dissolved gases.

. Relatively short open intervals of 10 feet so mixing likely minimized.

. Disadvantages associated with these samples:

. Mixture of domestic and commercial wells, so variable pumping rates.

. Median penetration of center of open interval into water table was 62.73 feet (not sampling close to the water table, potentially mixing).

. No tritium analyses.

. Depth to water (can affect tracer transport to water table):

.   Median: 13.39 feet

.   Mean: 20.13 feet

.   Min: 11.39 feet

.   Max: 51.39 feet

. Brief analysis:

.   The SF$_6$-based age gradient for these sites is shown in figure B70. The age gradient shows a general trend of increasing age with depth. Differences in screen length, recharge source/strength, aquifer heterogeneity, pumping stresses, and the position of the well within the flow system may cause some wells to deviate from the general pattern of increasing age with depth.

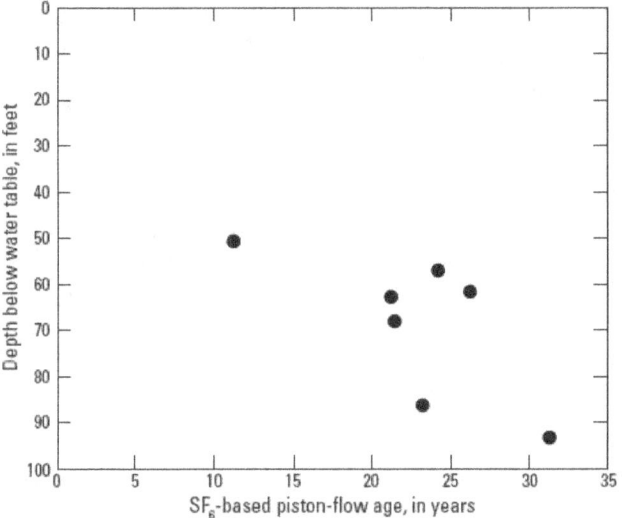

Figure B70.   SF$_6$-based age gradient for dated sites from the SUS2 network, LINJ Study Unit.

MISE LUSRC1, LUSRC2, and REFOT1

Samples from 31 sites in the MISE Study Unit were collected in 2006 for SF$_6$ (networks and, in parentheses, number of sites):

. LUSRC1 (23)

. LUSRC2 (7)

. REFOT1 (1)

The aquifer is composed of sand, silt, gravel, and clay of the Memphis Sand and Terrace deposits.

Major dissolved-gas data were available for 30 sites. Of these 30 sites, 12 were oxic and 18 were suboxic.

Age interpretations from tracer concentrations were made assuming that recharge elevation was equal to the elevation of the water table. Estimates of recharge temperature and excess air were based on major dissolved-gas data, with recharge temperature and excess air at suboxic sites being constrained using median excess air at oxic sites.

The raw tracer data, major dissolved-gas data, the ancillary chemical and well construction data that were used in the interpretations, and the piston-flow ages are presented in table B22.

. Advantages associated with these samples:

.   SF$_6$, as well as major dissolved gases.

.   Monitoring wells, so generally low pumping stress.

.   Relatively short open intervals ranging from 3.9 to 20 feet so mixing likely minimized.

.   Median penetration of center of open interval into water table was 15.84 feet (sampling close to the water table, potentially minimizing mixing).

. Depth to water (can affect tracer transport to water table):

.   Median: 32.39 feet

.   Mean: 43.72 feet

.   Min: 13.98 feet

.   Max: 97.28 feet

. Brief analysis:

.   The SF$_6$-based age gradient for these sites is shown in figure B71. The age gradient has a slight trend towards increasing age with depth. Differences in screen length, recharge source/strength, aquifer heterogeneity, pumping stresses, and the position of the well within the flow system may cause some wells to deviate from the general pattern of increasing age with depth.

The reconstructed ^{3}H plot for SF$_6$-based ages is shown in figure B72. The reconstruction shows evidence of unmixed, piston-flow transport for some samples, and mixing for other samples.

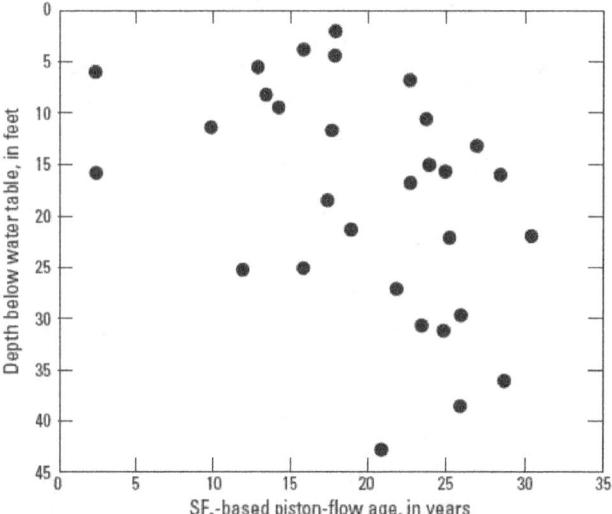

Figure B71.   SF$_6$-based age gradient for dated sites from the LUSRC1, LUSRC2, and REFOT1 networks, MISE Study Unit.

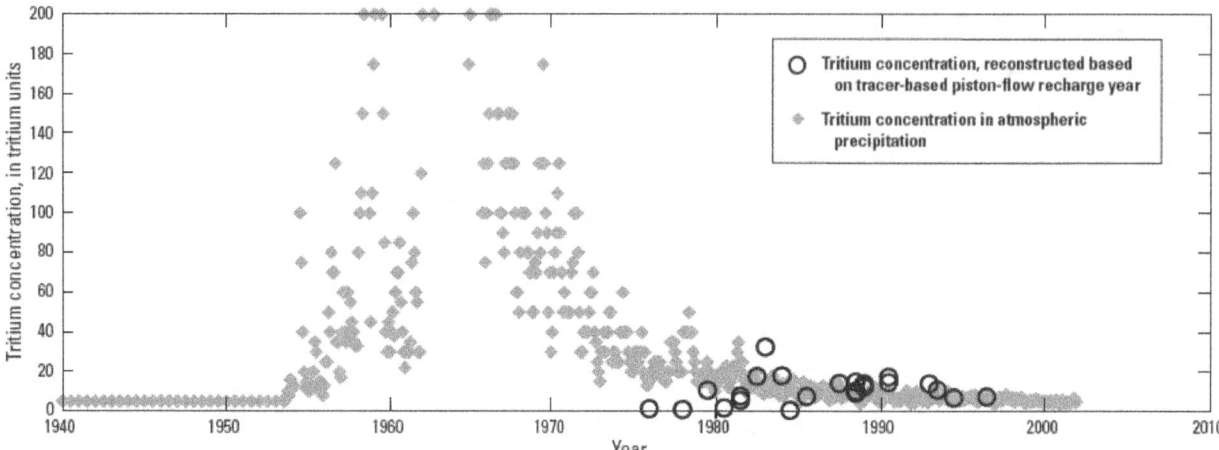

Figure B72.   Reconstructed tritium concentrations (using SF$_6$-based ages) and tritium in atmospheric precipitation, LUSRC1, LUSRC2, and REFOT1 networks, MISE Study Unit.

MISE SUS4

Samples from 16 sites in the MISE Study Unit were collected in 2007 for CFCs and ^{3}H/^{3}He (networks and, in parentheses, number of sites):

.   SUS4 (16)

The aquifer is composed of sand of the Memphis and Sparta Sands.

Age interpretations from tracer concentrations were made assuming that recharge elevation was equal to the elevation of the water table, that recharge temperature was equal to the mean annual air temperature +1°C, and that excess air concentrations were 2 cc STP/kg.

^{3}H/^{3}He ages were calculated for two sites (these two sites did not require a correction for terrigenic He), while five sites were not datable because tritium concentrations were too low, four sites were not datable because of fractionation, and samples from five sites were lost due to high pressure.

The raw tracer data, the ancillary chemical and well construction data that were used in the interpretations, and the piston-flow ages are presented in table B23.

.   Advantages associated with these samples:

.   Multiple tracers (CFCs and ^{3}H/^{3}He).

.   Disadvantages associated with these samples:

.   No major dissolved gases.

.   Mixture of domestic, irrigation, commercial, and public supply wells, so variable pumping rates.

.   Relatively large open intervals ranging from 5 to 75 feet so mixing likely.

.   Median penetration of center of open interval into water table was 178.18 feet (not sampling close to the water table, potentially mixing).

.   Depth to water (can affect tracer transport to water table):

.   Median: 35.87 feet

.   Mean: 49.65 feet

.   Min: 0.83 feet

.   Max: 128.12 feet

.   Brief analysis:

The CFC-based age gradient for these sites is shown in figure B73. The age gradient is typical of samples that are affected by CFC-degradation, and/or are at the limit of CFC-based dating. Three of the samples with slightly younger ages come from wells with slightly more oxic water than the rest of the samples, which would be consistent with degradation, however, many of the samples have extremely low tritium values, which would indicate very old ages that would be beyond the range of dating successfully with CFCs.

The reconstructed ^{3}H plot for CFC-based ages is shown in figure B74. The reconstruction shows evidence of unmixed, piston-flow transport for the three more oxic samples noted above, mixing for other samples, and ages beyond the range of dating using CFCs for others samples.

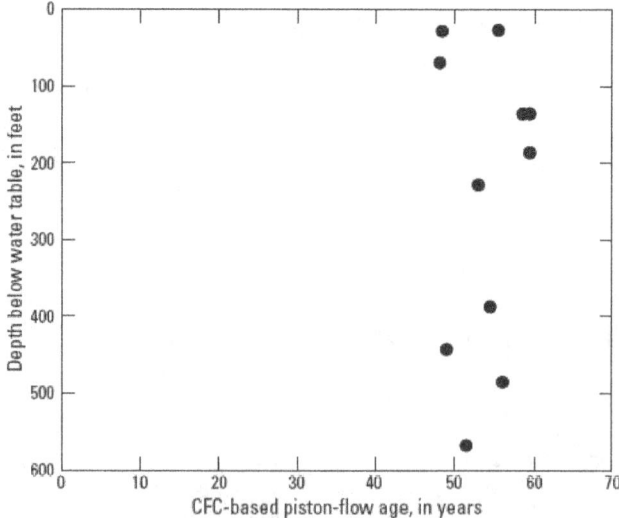

Figure B73.   CFC-based age gradient for dated sites from the SUS4 network, MISE Study Unit.

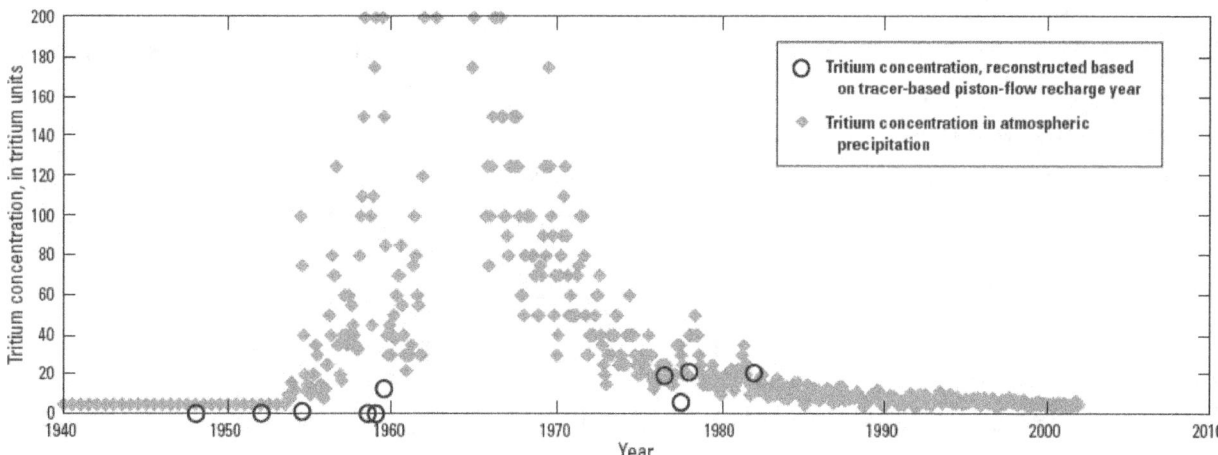

Figure B74. Reconstructed tritium concentrations (using CFC-based ages) and tritium in atmospheric precipitation, SUS4 network, MISE Study Unit.

NECB SUS3 and REFFO1

Samples from seven sites in the NECB Study Unit were collected in 2007 for CFCs (networks and, in parentheses, number of sites):

. SUS3 (5)

. REFFO1 (2)

The aquifer is composed of undifferentiated, stratified sand and gravel deposits.

Major dissolved-gas data were available for all seven sites. Of these seven sites, two were oxic and five were suboxic.

Age interpretations from tracer concentrations were made assuming that recharge elevation was equal to the elevation of the water table. Estimates of recharge temperature and excess air were based on major dissolved-gas data, with recharge temperature and excess air at suboxic sites being constrained using median excess air at oxic sites.

The raw tracer data, major dissolved-gas data, the ancillary chemical and well construction data that were used in the interpretations, and the piston-flow ages are presented in table B24.

. Advantages associated with these samples:

. CFCs, as well as major dissolved gases.

. Relatively short open intervals ranging from 1.3 to 20 feet so mixing likely minimized.

. Disadvantages associated with these samples:

. Mixture of monitoring and public supply wells, so variable pumping rates.

. Median penetration of center of open interval into water table was 47.04 feet (not sampling close to the water table, potentially mixing).

. Depth to water (can affect tracer transport to water table):

. Median: 13.83 feet

. Mean: 16.17 feet

. Min: 1.85 feet

. Max: 37.00 feet

. Brief analysis:

. Due to the suboxic conditions in this network, age-dating was not possible using CFCs.

NVBR LUSRC1, REFOT1, and SUS2

Samples from 13 sites in the NVBR Study Unit were collected in 2008 for CFCs and $^3H/^3He$ (networks and, in parentheses, number of sites):

. LUSRC1 (6)

. REFOT1 (2)

. SUS2 (5)

The aquifer is composed of valley-fill sand, gravel, clay, and silt deposits.

Major dissolved-gas data were available for 13 sites. Of these 13 sites, 8 were oxic and 5 were suboxic.

Age interpretations from tracer concentrations were made assuming that recharge elevation was equal to the elevation of the water table. Estimates of recharge temperature and excess air were based on major dissolved-gas data, with recharge temperature and excess air at suboxic sites being constrained using median excess air at oxic sites.

$^3H/^3He$ ages were calculated for six sites (four of the six sites required a correction for terrigenic helium), while three sites were not datable because tritium concentrations were too low, and two sites were not datable because of fractionation.

The raw tracer data, major dissolved-gas data, the ancillary chemical and well construction data that were used in the interpretations, and the piston-flow ages are presented in table B25.

. Advantages associated with these samples:

. Multiple tracers (CFCs and ^{3}H/^{3}He, as well as major dissolved gases).

. Disadvantages associated with these samples:

. Mixture of monitoring and public supply wells, so variable pumping rates.

. Relatively large open intervals ranging from 0 to 260 feet so mixing likely.

. Median penetration of center of open interval into water table was 24.81 feet (not sampling close to the water table, potentially mixing).

. Depth to water (can affect tracer transport to water table):

. Median: 44.73 feet

. Mean: 56.05 feet

. Min: 5.18 feet

. Max: 140.81 feet

. Brief analysis:

. The ^{3}H/^{3}He-based age gradient for these sites is shown in figure B75. The age gradient shows a general increase in age with depth if the oldest sample is included, despite being beyond the dating range for ^{3}H/^{3}He dating. If this sample is excluded, the age gradient is more scattered, with old ages at very shallow depths in monitoring wells. Differences in screen length, recharge source/strength, aquifer heterogeneity, pumping stresses, and the position of the well within the flow system may cause some wells to deviate from the general pattern of increasing age with depth.

The reconstructed ^{3}H plot for ^{3}H/^{3}He-based ages is shown in figure B76. The reconstruction shows evidence of unmixed, piston-flow transport for samples recharged since 1980, and mixing of pre- and post-bomb water for the sample with a piston-flow age of about 1970. Most of the samples have very low tritium concentrations, some of which preclude using the ^{3}H/^{3}He technique for dating.

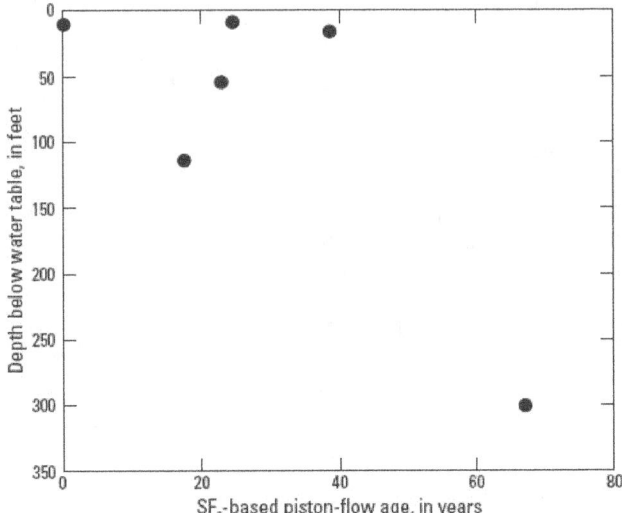

Figure B75. ^{3}H/^{3}He-based age gradient for dated sites from the LUSRC1, REFOT1, and SUS2 networks, NVBR Study Unit.

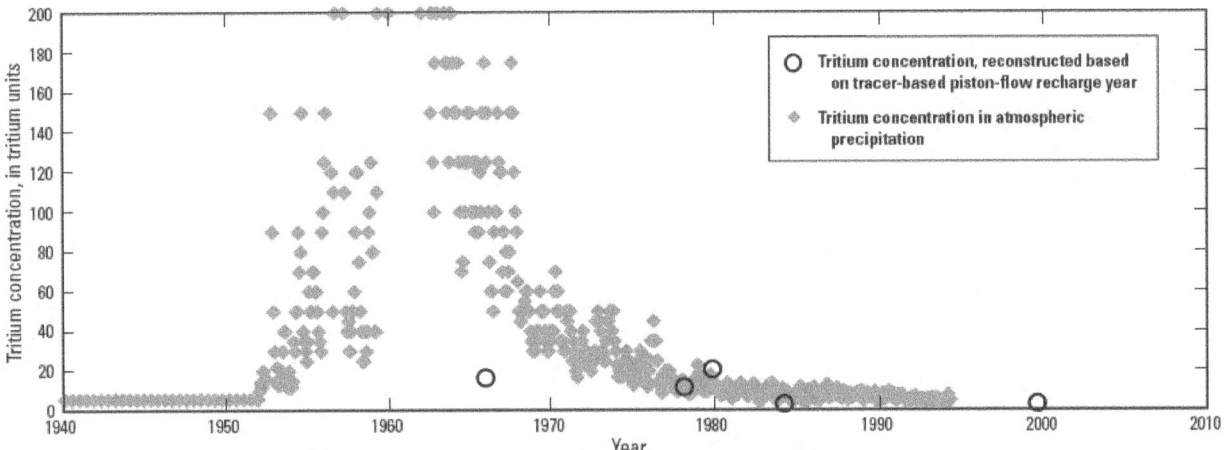

Figure B76. Reconstructed tritium concentrations (using ^{3}H/^{3}He-based ages) and tritium in atmospheric precipitation, LUSRC1, REFOT1, and SUS2 networks, NVBR Study Unit.

PODL LUSRC1

Samples from six sites in the PODL Study Unit were collected in 2007 for SF_6 and $^3H/^3He$ (networks and, in parentheses, number of sites):

.   LUSRC1 (5)

The aquifer is composed of fractured rock, with one well finished in limestone.

Major dissolved-gas data were available for five sites. Of these five sites, two were oxic and three were suboxic.

Age interpretations from tracer concentrations were made assuming that recharge elevation was equal to the elevation of the water table. Estimates of recharge temperature and excess air were based on major dissolved-gas data (PODL LUSRC1 and PODL REFFO1), with recharge temperature and excess air at suboxic sites being constrained using median excess air at oxic sites.

$^3H/^3He$ ages were calculated for three sites (two of the three sites required a correction for terrigenic helium), while two sites were not datable because of fractionation.

The raw tracer data, major dissolved-gas data, the ancillary chemical and well construction data that were used in the interpretations, and the piston-flow ages are presented in table B26.

.   Advantages associated with these samples:

.   Multiple tracers (SF_6 and $^3H/^3He$, as well as major dissolved gases).

.   Mostly monitoring wells, with one recreational well, so likely low pumping stress.

.   Disadvantages associated with these samples:

.   Relatively large open intervals ranging from 24 to 102 feet so mixing likely.

.   Median penetration of center of open interval into water table was 29.62 feet (not sampling close to the water table, potentially mixing).

.   Depth to water (can affect tracer transport to water table):

.   Median: 13.00 feet

.   Mean: 16.36 feet

.   Min: 6.83 feet

.   Max: 34.92 feet

.   Brief analysis:

.   The SF_6- and $^3H/^3He$-based age gradients for these sites are shown in figures B77 and B78. The age gradients show a similar pattern for both SF_6 and $^3H/^3He$ samples, with somewhat younger ages for SF_6. The wells in this network are from fractured rock and are not from a single flowpath, so it would not be expected to have a clearly defined age gradient, particularly for so few wells.

The reconstructed ^3H plots for SF_6- and $^3H/^3He$-based ages are shown in figures B79 and B80. The reconstructions are similar and show evidence of unmixed, piston-flow transport.

The $^3H/^3He$- versus SF_6-based age comparison for this network is shown in figure B81. The age comparison is limited by the low number of samples and scattered values.

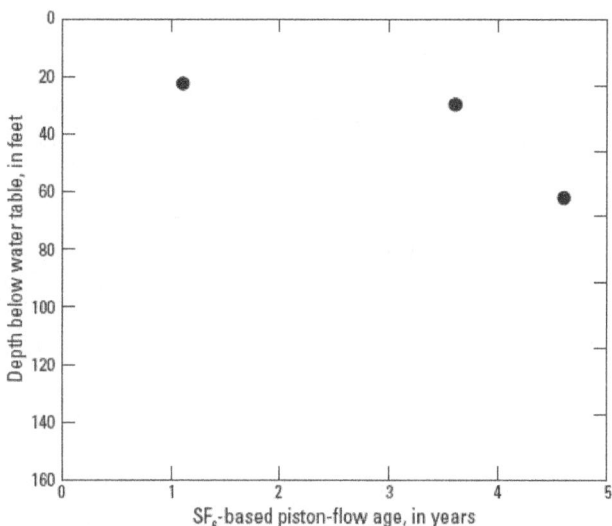

Figure B77. SF_6-based age gradient for dated sites from the LUSRC1 network, PODL Study Unit.

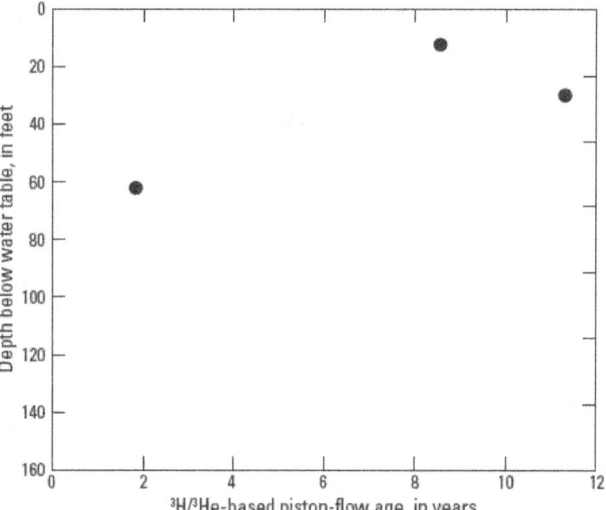

Figure B78. $^3H/^3He$-based age gradient for dated sites from the LUSRC1 network, PODL Study Unit.

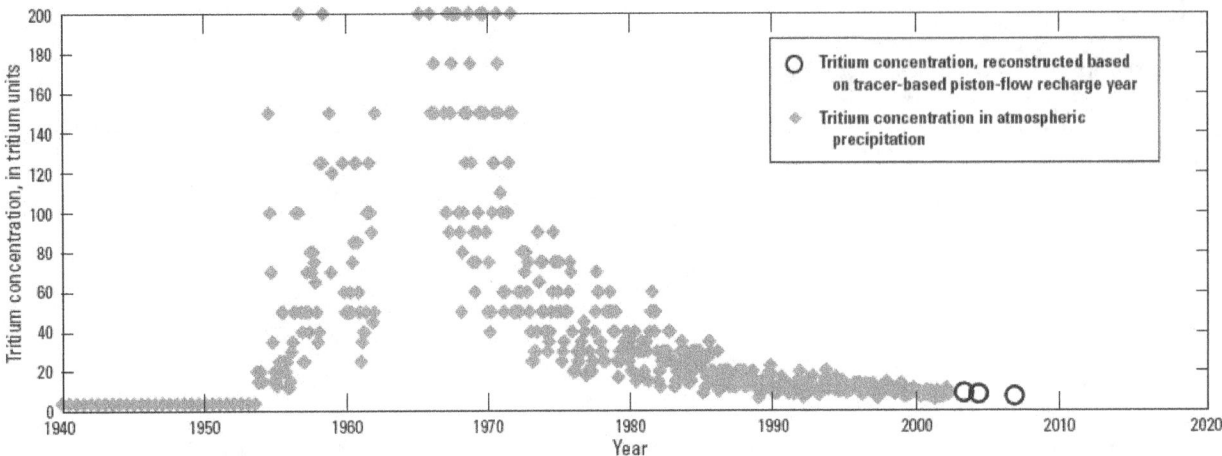

Figure B79.   Reconstructed tritium concentrations (using SF$_6$-based ages) and tritium in atmospheric precipitation, LUSRC1 network, PODL Study Unit.

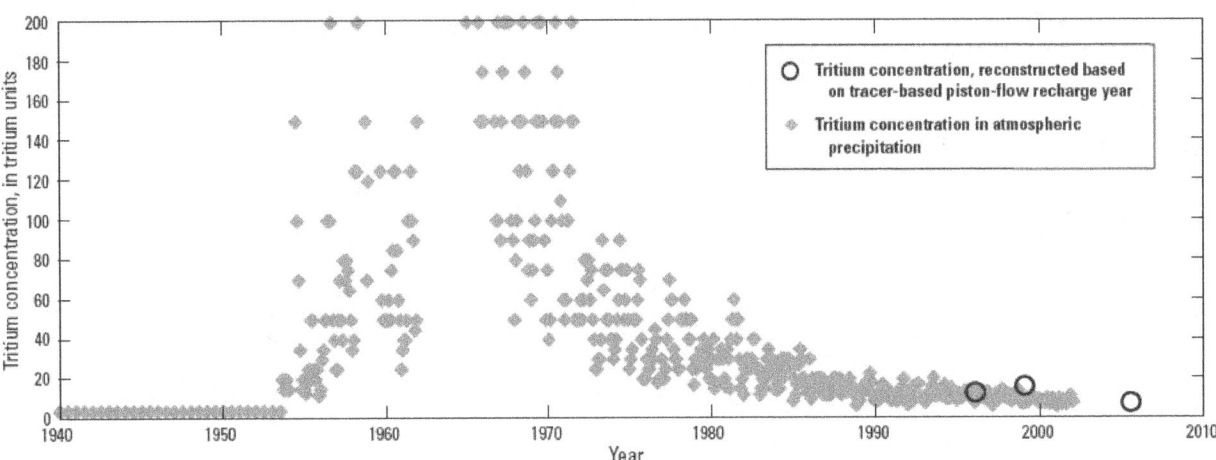

Figure B80.   Reconstructed tritium concentrations (using ^{3}H/^{3}He-based ages) and tritium in atmospheric precipitation, LUSRC1 network, PODL Study Unit.

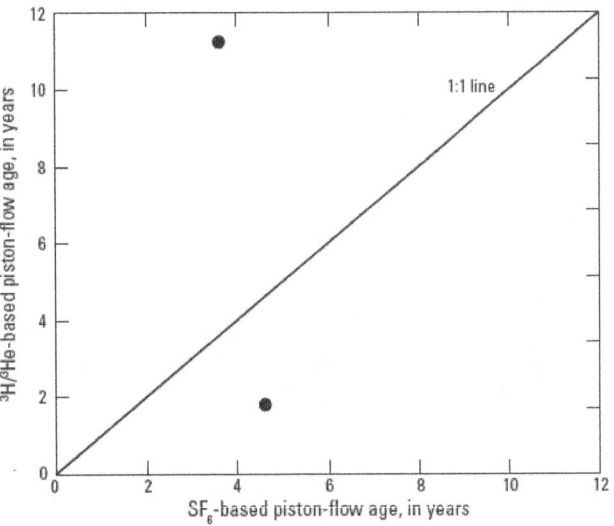

Figure B81. ^{3}H/^{3}He- versus SF$_6$-based age comparison, LUSRC1 network, PODL Study Unit.

PODL POTOLUSAG1 and PODLREFFO1

Samples from six sites in the PODL Study Unit were collected in 2007 for SF$_6$ and ^{3}H/^{3}He (networks and, in parentheses, number of sites):

.   POTOLUSAG1 (5)

.   PODLREFFO1 (1)

The aquifer is composed of limestone and dolomite.

Major dissolved-gas data were available for six sites. Of these six sites, five were oxic and one was suboxic.

Age interpretations from tracer concentrations were made assuming that recharge elevation was equal to the elevation of the water table. Estimates of recharge temperature and excess air were based on major dissolved-gas data, with recharge temperature and excess air at suboxic sites being constrained using median excess air at oxic sites.

^{3}H/^{3}He ages were calculated for four sites (two of the four sites required a correction for terrigenic helium), while one site was lost due to high pressure.

The raw tracer data, major dissolved-gas data, the ancillary chemical and well construction data that were used in the interpretations, and the piston-flow ages are presented in table B27.

. Advantages associated with these samples:

   . Multiple tracers (SF_6 and $^3H/^3He$, as well as major dissolved gases).

   . Mixture of domestic and recreation wells, so likely low pumping stress.

. Disadvantages associated with these samples:

   . Relatively large open intervals ranging from 17 to 109 feet so mixing likely.

   . Median penetration of center of open interval into water table was 62.86 feet (not sampling close to the water table, potentially mixing).

. Depth to water (can affect tracer transport to water table):

. Median: 42.15 feet
. Mean: 55.43 feet
. Min: 29.50 feet
. Max: 40.00 feet

. Brief analysis:

The SF_6- and $^3H/^3He$-based age gradients for these sites are shown in figures B82 and B83. The age gradients show a similar pattern for both SF_6 and $^3H/^3He$ samples, with somewhat younger ages for SF_6. The wells in this network are from karst and are not from a single flowpath, so it would not be expected to have a clearly defined age gradient, particularly for so few wells.

The reconstructed ^3H plots for SF_6- and $^3H/^3He$-based ages are shown in figures B84 and B85. The reconstructions are similar, which may simply be the result from the samples plotting in the relatively flat portion of the ^3H reconstruction.

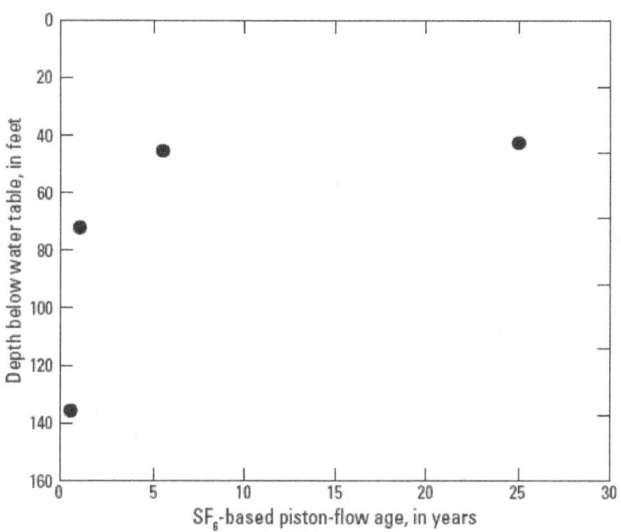

Figure B82. SF_6-based age gradient for dated sites from the POTOLUSAG1 and PODLREFF01 networks, PODL Study Unit.

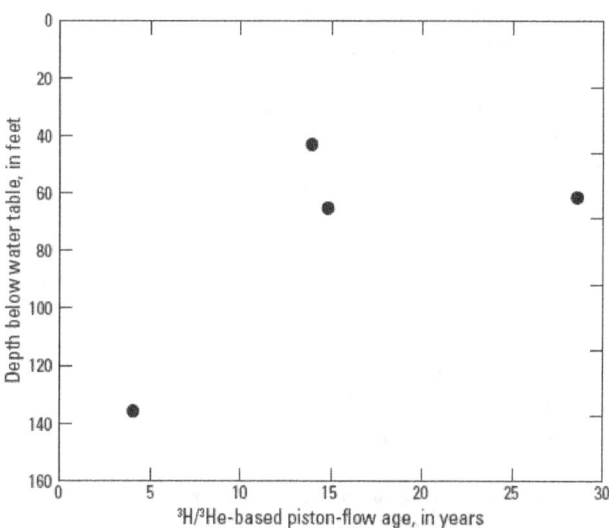

Figure B83. $^3H/^3He$-based age gradient for dated sites from the POTOLUSAG1 and PODLREFF01 networks, PODL Study Unit.

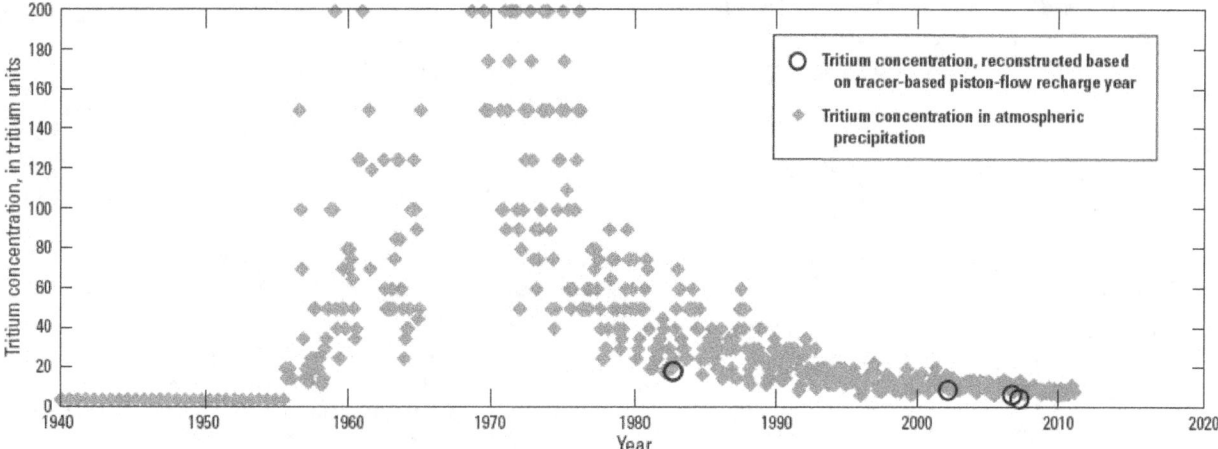

Figure B84.   Reconstructed tritium concentrations (using SF_6-based ages) and tritium in atmospheric precipitation, POTOLUSAG1 and PODLREFF01 networks, PODL Study Unit.

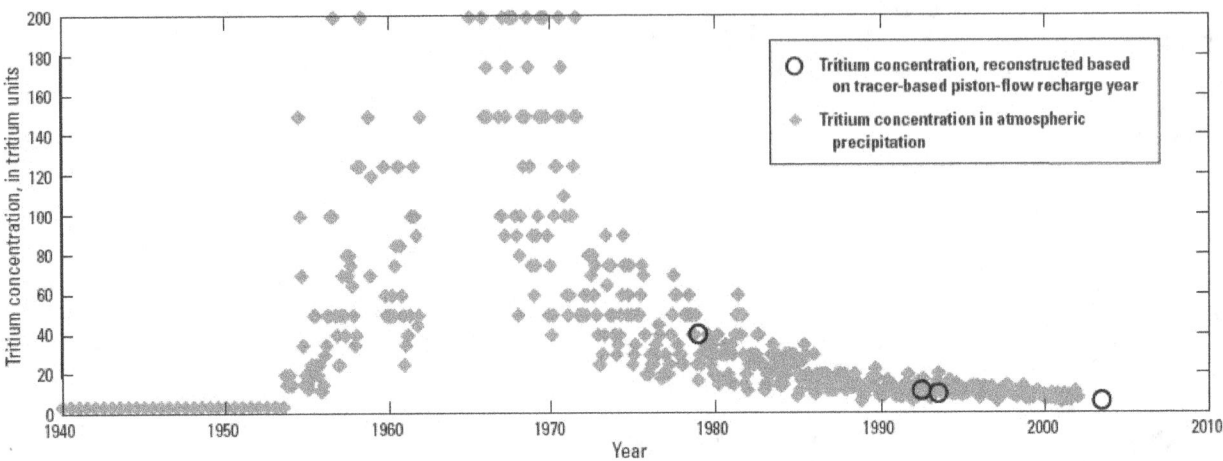

Figure B85.   Reconstructed tritium concentrations (using $^3H/^3He$-based ages) and tritium in atmospheric precipitation, POTOLUSAG1 and PODLREFFO1 networks, PODL Study Unit.

PUGT FPSCR1 and LUSCR1

Samples from 20 sites in the PUGT Study Unit were collected in 2007 for SF_6 and $^3H/^3He$ (networks and, in parentheses, number of sites):

.   FPSCR1 (11)

.   LUSCR1 (9)

The aquifer is composed of glacial sands, gravel, and clay.

Major dissolved-gas data were available for 20 sites. Of these 20 sites, 7 were oxic and 13 were suboxic.

Age interpretations from tracer concentrations were made assuming that recharge elevation was equal to the elevation of the water table. Estimates of recharge temperature and excess air were based on major dissolved-gas data, with recharge temperature and excess air at suboxic sites being constrained using median excess air at oxic sites.

$^3H/^3He$ ages were calculated for eight sites (only one of the eight sites required a correction for terrigenic helium).

The raw tracer data, major dissolved-gas data, the ancillary chemical and well construction data that were used in the interpretations, and the piston-flow ages are presented in table B28.

.   Advantages associated with these samples:

. SF_6 and $^3H/^3He$, as well as major dissolved gases.

.   Monitoring wells, so low pumping stress.

.   Relatively short open intervals ranging from 1.98 to 54.08 feet so mixing likely minimized.

.   Median penetration of center of open interval into water table was 22.97 feet (sampling close to the water table, potentially minimizing mixing).

.   Disadvantages associated with these samples:

.   Suboxic conditions.

.   Depth to water (can affect tracer transport to water table):

.   Median: 9.42 feet

.   Mean: 8.93 feet

.   Min: 3.95 feet

.   Max: 18.32 feet

.   Brief analysis:

.   The SF_6- and $^3H/^3He$-based age gradients for these sites are shown in figures B86 and B87. The SF_6-based age gradient shows a great deal of scatter, which may be the result of gas stripping due to the suboxic conditions. The $^3H/^3He$-based age gradient shows a general increase in age with depth. Differences in screen length, recharge source/strength, aquifer heterogeneity, pumping stresses, and the position of the well within the flow system may cause some wells to deviate from the general pattern of increasing age with depth.

The reconstructed ^3H plots for SF_6- and $^3H/^3He$-based ages are shown in figures B88 and B89. The reconstructions show evidence of unmixed, piston-flow transport for most samples, and mixing or dispersion for other samples in the SF_6 reconstruction. Some of these samples may have been affected by gas stripping due to elevated methane (3 of the 4 samples have measurable methane). For those samples, the SF_6 ages would in fact be younger and would fall on the reconstructed tritium input curve.

The $^3H/^3He$- versus SF_6-based age comparison for this network is shown in figure B90. Several samples may be affected by gas stripping and SF_6 loss, which could explain the discrepancy in ages.

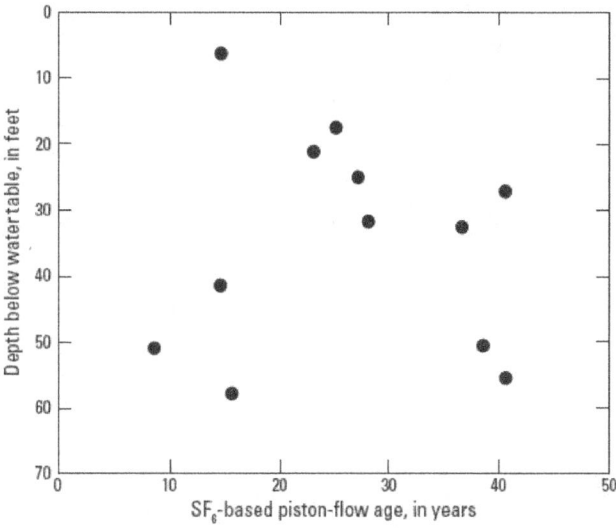

Figure B86. SF$_6$-based age gradient for dated sites from the FPSCR1 and LUSCR1 networks, PUGT Study Unit.

Figure B87. ^{3}H/^{3}He-based age gradient for dated sites from the FPSCR1 and LUSCR1 networks, PUGT Study Unit.

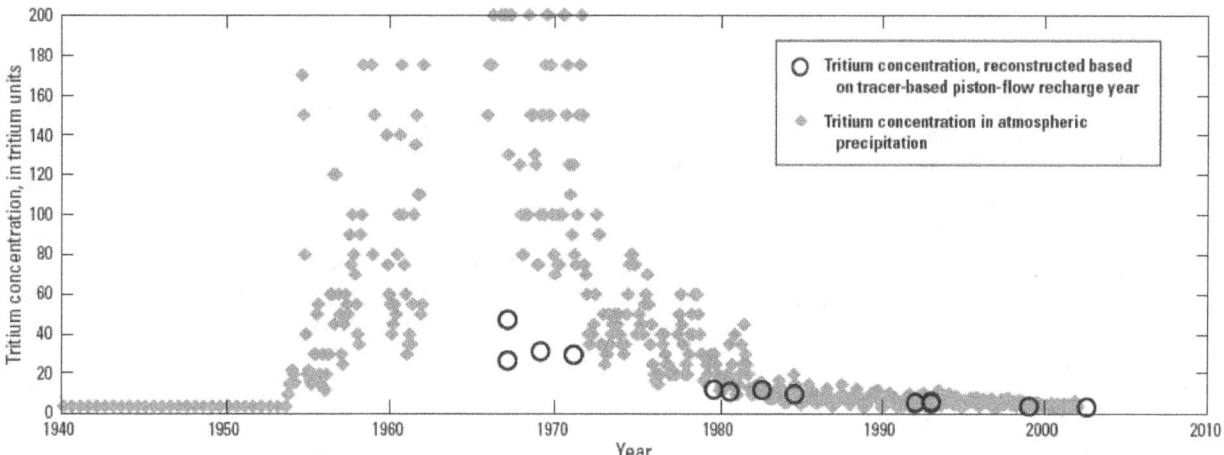

Figure B88. Reconstructed tritium concentrations (using SF$_6$-based ages) and tritium in atmospheric precipitation, FPSCR1 and LUSCR1 networks, PUGT Study Unit.

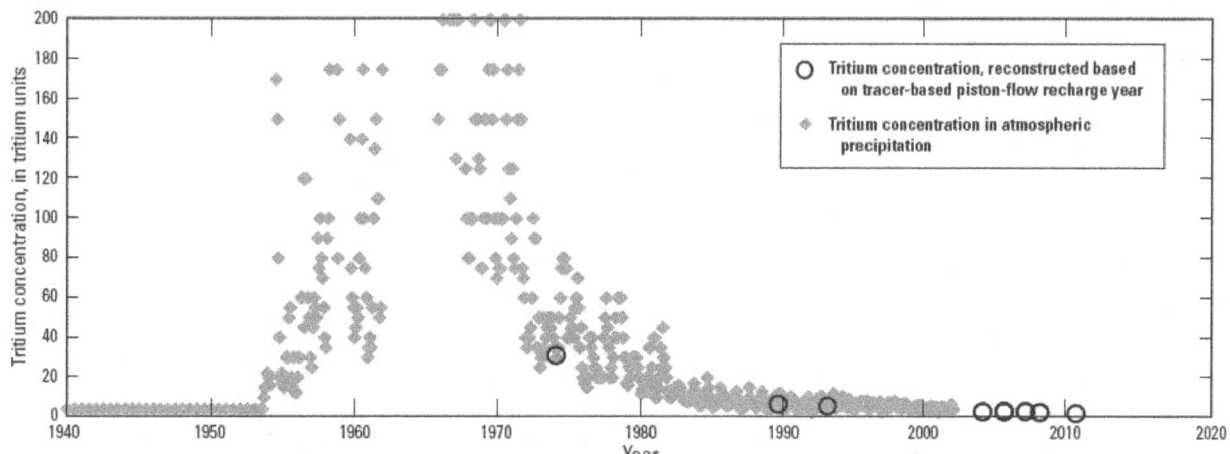

Figure B89. Reconstructed tritium concentrations (using ^{3}H/^{3}He-based ages) and tritium in atmospheric precipitation, FPSCR1 and LUSCR1 networks, PUGT Study Unit.

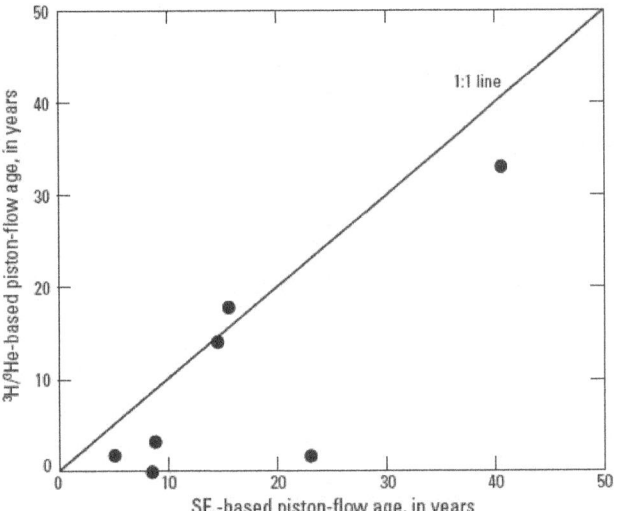

Figure B90. $^3H/^3He$- versus SF_6-based age comparison, FPSCR1, and LUSCR1 networks, PUGT Study Unit.

RIOG LUSAG1

Samples from 30 sites in the RIOG Study Unit were collected in 2006 for CFCs and 5 sites were repeated in 2008 for $^3H/^3He$ (network and, in parentheses, number of sites):

. LUSAG1 (CFCs, 30; $^3H/^3He$, 5)

The aquifer is composed of alluvial sand and silt.

Major dissolved-gas data were available for 22 sites from the 2006 sampling. Of these 22 sites, 1 was oxic and 21 were suboxic, and 3 were degassed. Excess air concentrations were elevated and difficult to constrain. Recharge temperatures ranged from 11.2 to 25.5°C and bracketed the 17.7°C determined from MAAT +1°C.

Age interpretations from tracer concentrations were made assuming that recharge elevation was equal to the elevation of the water table, that recharge temperature was equal to mean annual air temperature +1°C, and that excess air concentrations were 2 cc STP/kg.

$^3H/^3He$ age was calculated for only one site (this one site did not require a correction for terrigenic He), while three sites were not datable because of fractionation, and the sample from one site was lost due to high pressure.

The raw tracer data, major dissolved-gas data, the ancillary chemical and well construction data that were used in the interpretations, and the piston-flow ages are presented in table B29.

. Advantages associated with these samples:

. Multiple tracers (CFCs and $^3H/^3He$, as well as major dissolved gases).

. Monitoring wells, therefore low pumping stress.

Relatively short open intervals ranging from 3.34 to 10 feet so mixing likely minimized.

Median penetration of center of open interval into water table was 3.76 feet (sampling close to the water table, potentially minimizes mixing).

. Disadvantages associated with these samples:

Suboxic conditions and degassing.

Only five tritium values.

. Depth to water (can affect tracer transport to water table):

Median: 11.68 feet

Mean: 12.01 feet

Min: 3.10 feet

Max: 18.22 feet

. Brief analysis:

The CFC-based age gradient for these sites is shown in figure B91. The age gradient shows a great deal of scatter and a shift toward older ages for very shallow depths as would be expected for a network with suboxic conditions and CFC degradation. In fact, most of the CFC samples plotted in figure B79 do not have methane measurements and were therefore left as fixed ages, when in fact, they likely have elevated methane concentrations and should simply be bracketed in terms of ages as was done for all of the other samples. The CFC data for this network provide an excellent example of the effects of degradation processes on CFC concentrations. Table B29 shows a consistent pattern of CFC-11 ages being the oldest (most degraded), CFC-113 ages being older than (or similar to) CFC-12 ages, and CFC-12 ages being the youngest.

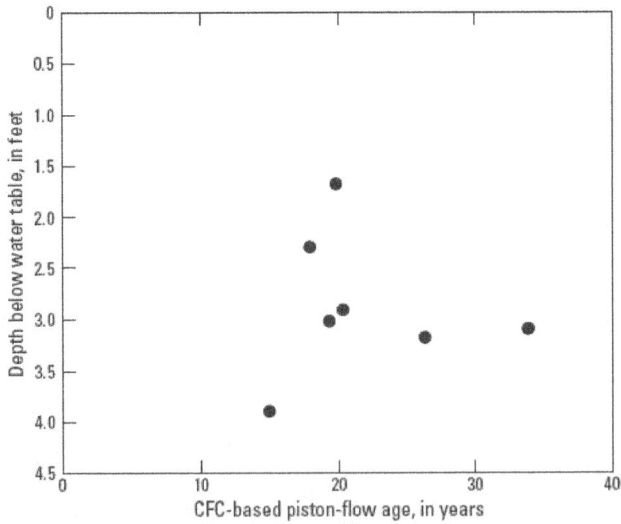

Figure B91.   CFC-based age gradient for dated sites from the LUSAG1 network, RIOG Study Unit.

RIOG LUSCR1

Samples from 58 sites in the RIOG Study Unit were collected in 2007 and 2008 for CFCs and ^{3}H/^{3}He (networks and, in parentheses, number of sites):

.   LUSCR1 (CFCs, 58; ^{3}H/^{3}He, 29)

The aquifer is composed of sand, gravel, and silt of the Rio Grande aquifer system.

Major dissolved-gas data were available for 16 sites. Of these 16 sites, 13 were oxic, and one of the suboxic sites had excess air concentrations below the average of the oxic sites and therefore did not require any corrections. The dissolved-gas recharge temperatures were clustered in two groups with one group having temperatures significantly above the other dissolved-gas samples or the MAAT +1°C estimates.

Age interpretations from tracer concentrations were made assuming that recharge elevation was equal to the elevation of the water table, that recharge temperature was equal to mean annual air temperature +1°C, and that excess air concentrations were 2 cc STP/kg.

^{3}H/^{3}He ages were calculated for 15 sites (only 2 of the 15 sites required a correction for terrigenic helium), while 5 sites were not datable because of fractionation, and samples from 9 sites were lost due to high pressure.

The raw tracer data, major dissolved-gas data, the ancillary chemical and well construction data that were used in the interpretations, and the piston-flow ages are presented in table B30.

.   Advantages associated with these samples:

.   Multiple tracers (CFCs and ^{3}H/^{3}He, as well as major dissolved gases).

.   Monitoring wells, therefore low pumping stress.

.   Relatively short open intervals ranging from 2.46 to 10 feet so mixing likely minimized.

.   Median penetration of center of open interval into water table was 11.45 feet (sampling close to the water table, potentially minimizing mixing).

.   Disadvantages associated with these samples:

.   None.

.   Depth to water (can affect tracer transport to water table):

.   Median: 18.68 feet

.   Mean: 18.84 feet

.   Min: 3.64 feet

.   Max: 61.11 feet

.   Brief analysis:

.   The CFC- and ^{3}H/^{3}He-based age gradients for these sites are shown in figures B92 and B93. The age gradients show a great deal of scatter with a shift toward older ages for the CFC-based age gradient. Differences in screen length, recharge source/strength,

aquifer heterogeneity, pumping stresses, and the position of the well within the flow system may cause some wells to deviate from the general pattern of increasing age with depth.

The reconstructed ^{3}H plots for CFC- and ^{3}H/^{3}He-based ages are shown in figures B94 and B95. The reconstructions show evidence of unmixed, piston-flow transport for most samples. The one old sample in the ^{3}H/^{3}He reconstruction has a very low tritium concentration and is likely affected by mixing of pre- and post-bomb water.

The ^{3}H/^{3}He- versus CFC-based age comparison for this network is shown in figure B96. The age comparison shows the shift in CFC-based ages toward older ages as compared to the ^{3}H/^{3}He-based ages.

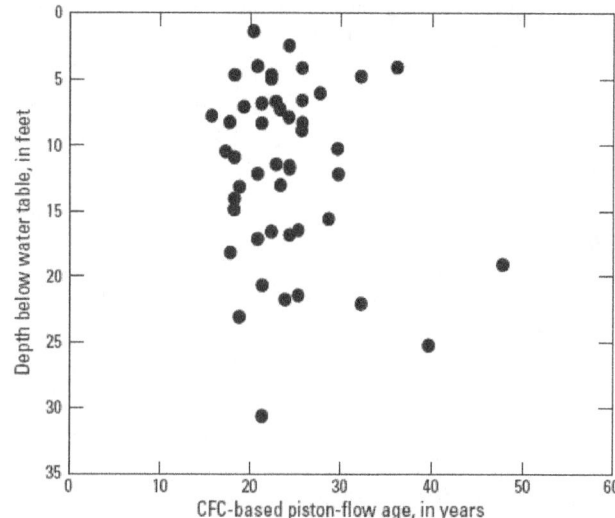

Figure B92.   CFC-based age gradient for dated sites from the LUSCR1 network, RIOG Study Unit.

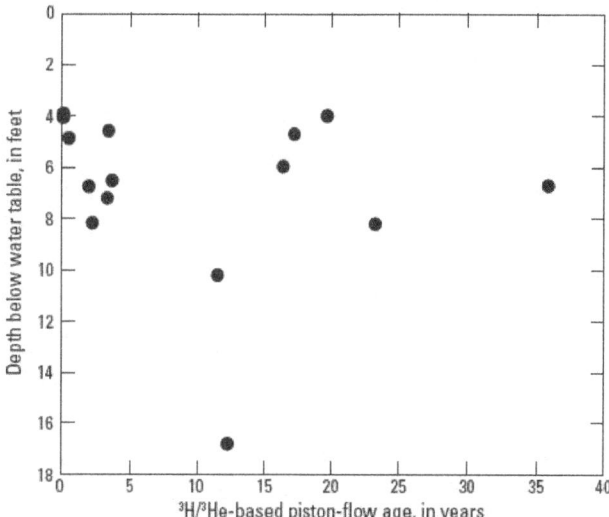

Figure B93. ^{3}H/^{3}He-based age gradient for dated sites from the LUSCR1 network, RIOG Study Unit.

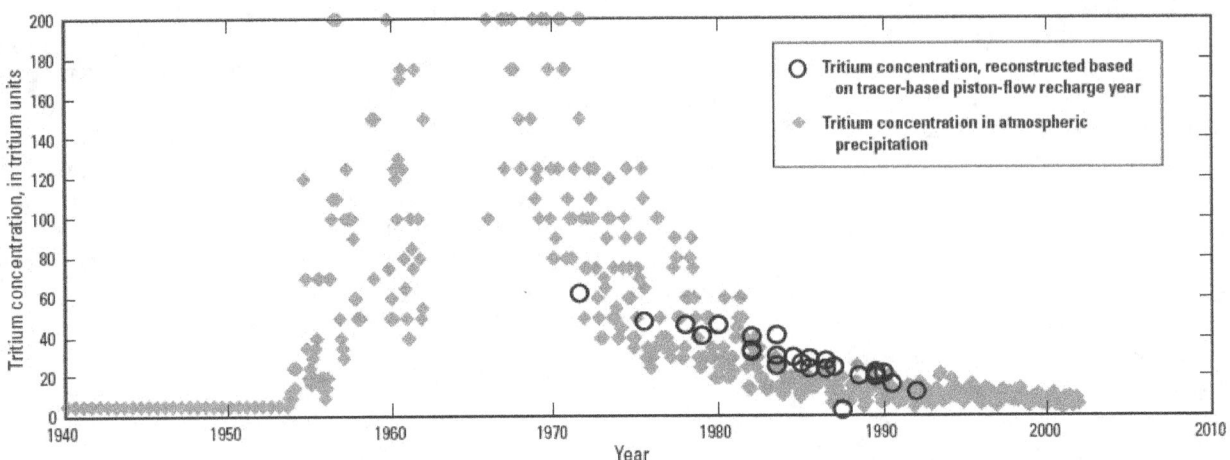

Figure B94.   Reconstructed tritium concentrations (using CFC-based ages) and tritium in atmospheric precipitation, LUSCR1 network, RIOG Study Unit.

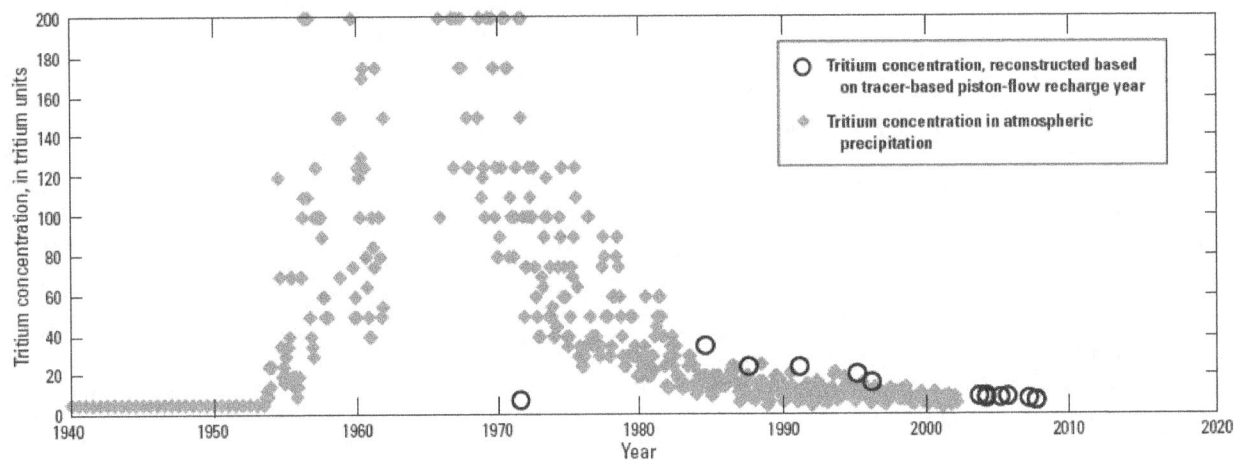

Figure B95.   Reconstructed tritium concentrations (using ^{3}H/^{3}He-based ages) and tritium in atmospheric precipitation, LUSCR1 network, RIOG Study Unit.

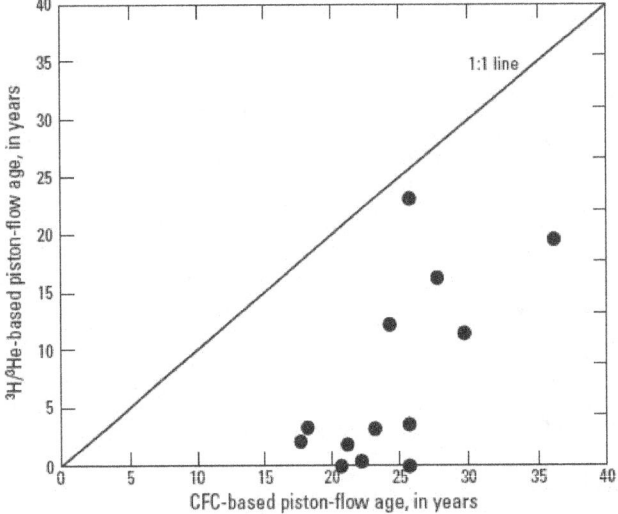

Figure B96. ^{3}H/^{3}He- versus CFC-based age comparison, LUSCR1 network, RIOG Study Unit.

RIOG LUSRC1

Samples from seven sites in the RIOG Study Unit were collected in 2008 for ^{3}H/^{3}He (networks and, in parentheses, number of sites):

. LUSRC1 (7)

The aquifer is composed of sands of the Rio Grande aquifer system.

Major dissolved-gas data were available for seven sites. Of these seven sites, two were oxic and five were suboxic.

Age interpretations from tracer concentrations were made assuming that recharge elevation was equal to the elevation of the water table, that recharge temperature was equal to mean annual air temperature +1°C, and that excess air concentrations were 2 cc STP/kg.

^{3}H/^{3}He ages were calculated for five sites (two of the five sites required a correction for terrigenic helium), while two sites were not datable because of fractionation.

The raw tracer data, major dissolved-gas data, the ancillary chemical and well construction data that were used in the interpretations, and the piston-flow ages are presented in table B31.

. Advantages associated with these samples:

  . ^{3}H/^{3}He, as well as major dissolved gases.

  . Monitoring wells, therefore low pumping stress.

  . Relatively short open intervals ranging from 4.14 to 10 feet so mixing likely minimized.

  . Median penetration of center of open interval into water table was 4.81 feet (sampling close to the water table, potentially minimizing mixing).

. Disadvantages associated with these samples:

  . Suboxic conditions.

. Depth to water (can affect tracer transport to water table):

  . Median: 8.18 feet

  . Mean: 12.26 feet

  . Min: 4.18 feet

  . Max: 23.86 feet

. Brief analysis:

  . The ^{3}H/^{3}He-based age gradient for these sites is shown in figure B97. The age gradient shows little structure, likely as a result of the narrow range in depths for these wells.

The reconstructed ^{3}H plot for ^{3}H/^{3}He-based ages is shown in figure B98. The reconstruction shows evidence of unmixed, piston-flow transport.

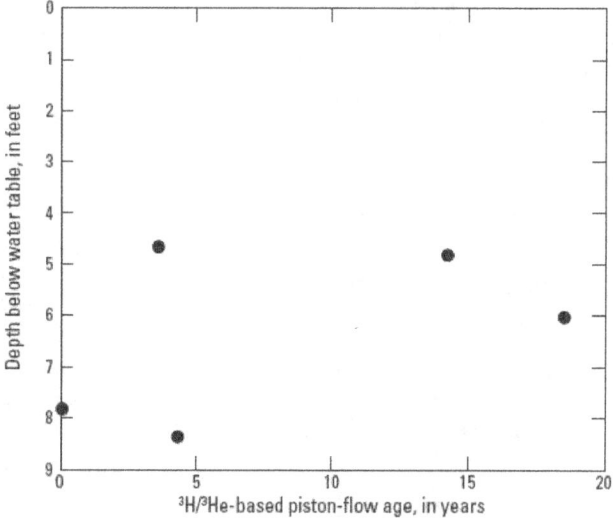

Figure B97. ^{3}H/^{3}He-based age gradient for dated sites from the LUSRC1 network, RIOG Study Unit.

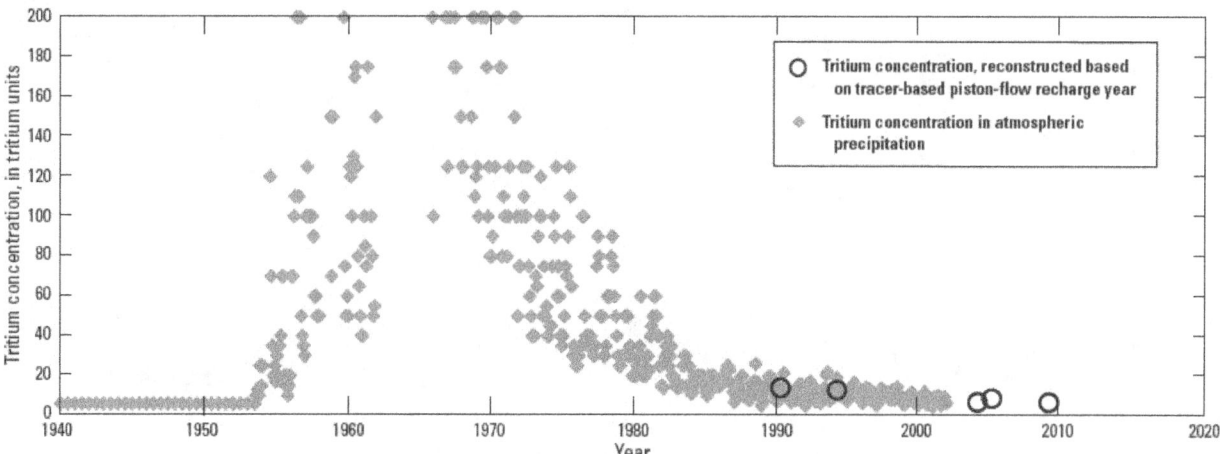

Figure B98.   Reconstructed tritium concentrations (using ^{3}H/^{3}He-based ages) and tritium in atmospheric precipitation, LUSRC1 network, RIOG Study Unit.

SACR LUSCR1

Samples from 5 sites in the SACR Study Unit were collected in 2006 for CFCs and SF_6 (networks and, in parentheses, number of sites):

. LUSCR1 (5)

The aquifer is composed of alluvial sand, gravel, silt and clay of the Central Valley aquifer system.

Major dissolved-gas data were available for all five sites. Of these five sites, one was oxic and four were suboxic.

Age interpretations from tracer concentrations were made assuming that recharge elevation was equal to the elevation of the water table, that recharge temperature was equal to mean annual air temperature +1°C, and that excess air concentrations were 2 cc STP/kg.

The raw tracer data, major dissolved-gas data, the ancillary chemical and well construction data that were used in the interpretations, and the piston-flow ages are presented in table B32.

. Advantages associated with these samples:

. Multiple tracers (CFCs and SF_6, as well as major dissolved gases).

. Monitoring wells, therefore low pumping stress.

. Relatively short open intervals ranging from 5 to 10 feet so mixing likely minimized.

. Disadvantages associated with these samples:

. Median penetration of center of open interval into water table was 26.04 feet (sampling close to the water table, potentially minimizing mixing).

. Depth to water (can affect tracer transport to water table):

. Median: 3.73 feet

. Mean: 5.05 feet

. Min: 1.67 feet

. Max: 13.05 feet

. Brief analysis:

The SF_6-based age gradient for these sites is shown in figure B99. The age gradient shows little structure, likely as a result of the narrow range in depths for these wells.

The reconstructed ^3H plot for $^3H/^3He$-based ages is shown in figure B100. The samples generally plot above the ^3H reconstruction suggesting that the samples may be of mixed age resulting from heterogeneity of aquifer sediments.

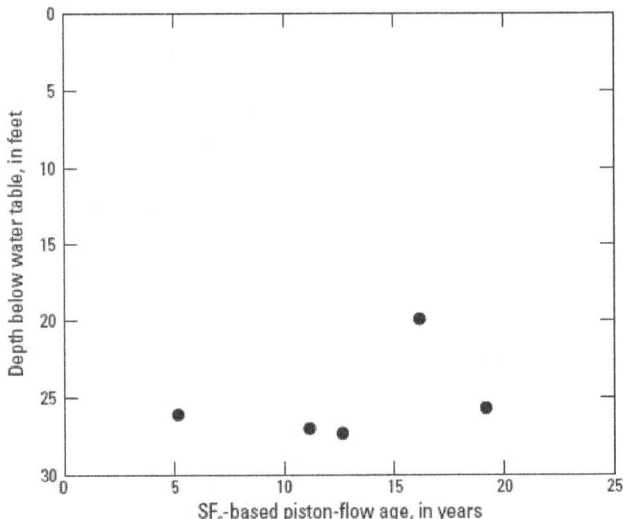

Figure B99. SF_6-based age gradient for dated sites from the LUSRC1 network, SACR Study Unit.

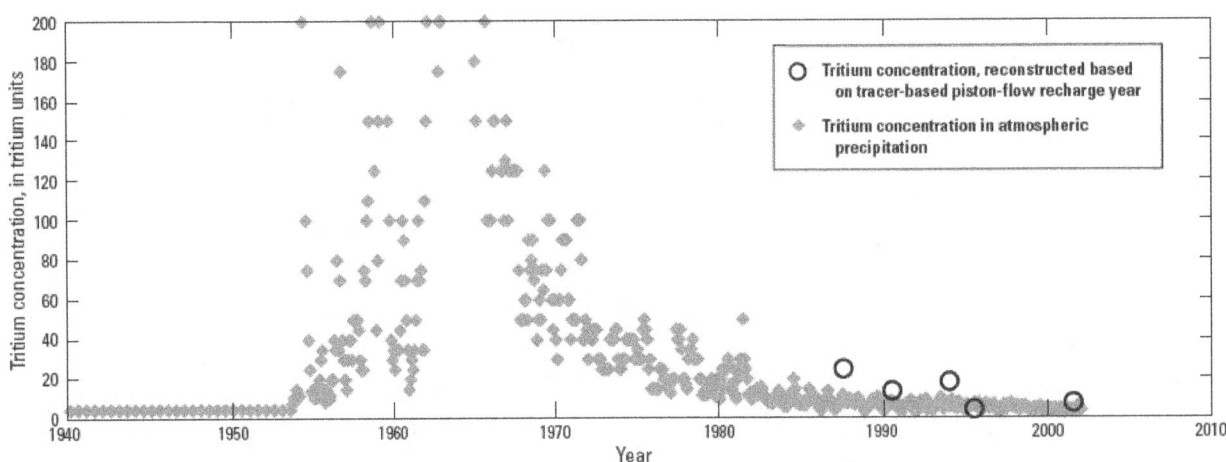

Figure B100.   Reconstructed tritium concentrations (using SF_6-based ages) and tritium in atmospheric precipitation, LUSRC1 network, SACR Study Unit.

SACR LUSRC1

Samples from five sites in the SACR Study Unit were collected in 2006 for CFCs and SF_6 (networks and, in parentheses, number of sites):

LUSRC1 (CFCs, 5; SF_6, 4)

The aquifer is composed of alluvial sand, gravel, silt and clay of the Central Valley aquifer system.

Major dissolved-gas data were available for all five sites. Of these five sites, three were oxic and two were suboxic, and one was degassed.

Age interpretations from tracer concentrations were made assuming that recharge elevation was equal to the elevation of the water table. Estimates of recharge temperature and excess air were based on major dissolved-gas data, with recharge temperature and excess air at suboxic sites being constrained using median excess air at oxic sites.

The raw tracer data, major dissolved-gas data, the ancillary chemical and well construction data that were used in the interpretations, and the piston-flow ages are presented in table B33.

. Advantages associated with these samples:

. Multiple tracers (CFCs and SF_6, as well as major dissolved gases).

. Monitoring wells, therefore low pumping stress.

. Relatively short open intervals of 10 feet so mixing likely minimized.

. Median penetration of center of open interval into water table was 18.88 feet (sampling close to the water table, potentially minimizing mixing).

. Disadvantages associated with these samples:

. No tritium analyses.

. Depth to water (can affect tracer transport to water table):

. Median: 19.53 feet

. Mean: 42.11 feet

. Min: 4.22 feet

. Max: 138.15 feet

. Brief analysis:

. The SF_6-based age gradient for these sites is shown in figure B101. It is difficult to ascertain the age structure for this study unit with only four samples, but the shallow wells were all younger than the deepest. Differences in screen length, recharge source/strength, aquifer heterogeneity, pumping stresses, and the position of the well within the flow system may cause some wells to deviate from the general pattern of increasing age with depth.

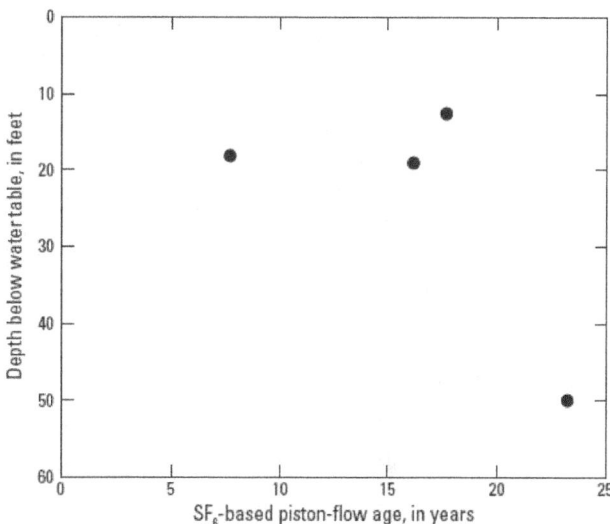

Figure B101. SF_6-based age gradient for dated sites from the LUSRC1 network, SACR Study Unit.

SACR SUS1

Samples from 26 sites in the SACR Study Unit were collected in 2008 for CFCs, SF_6, and $^3H/^3He$ (networks and, in parentheses, number of sites):

. SUS1 (CFCs, 26; SF_6, 26; $^3H/^3He$, 21)

The aquifer is composed of alluvial sand, gravel, silt and clay of the Central Valley aquifer system.

Major dissolved-gas data were available for 26 sites. Of these 26 sites, 18 were oxic and 8 were suboxic.

Age interpretations from tracer concentrations were made assuming that recharge elevation was equal to the elevation of the water table. Estimates of recharge temperature and excess air were based on major dissolved-gas data, with recharge temperature and excess air at suboxic sites being constrained using median excess air at oxic sites.

$^3H/^3He$ ages were calculated for twelve sites (5 of the 12 sites required a correction for terrigenic helium), while 1 site was not datable because tritium was too low, 2 sites were not datable because of fractionation, and samples from 6 sites were lost due to high pressure.

The raw tracer data, major dissolved-gas data, the ancillary chemical and well construction data that were used in the interpretations, and the piston-flow ages are presented in table B34.

. Advantages associated with these samples:

. Multiple tracers (CFCs, SF_6, and $^3H/^3He$, as well as major dissolved gases).

. Disadvantages associated with these samples:

  Mixture of domestic and irrigation wells, so variable
  pumping rates.

  Relatively large open intervals ranging from
  11 to 133 feet so mixing likely.

  Median penetration of center of open interval into
  water table was 68.19 feet (not sampling close to the
  water table, potentially mixing).

. Depth to water (can affect tracer transport to water table):

  Median: 29.43 feet

  Mean: 43.78 feet

  Min: 3.34 feet

  Max: 158.26 feet

. Brief analysis:

  The CFC-, SF$_6$-, and ^{3}H/^{3}He-based age gradients for
  these sites are shown in figures B102, B103, and B104.
  The age gradients do not show any particular structure.
  Differences in screen length, recharge source/strength,
  aquifer heterogeneity, pumping stresses, and the
  position of the well within the flow system may cause
  some wells to deviate from the general pattern of
  increasing age with depth.

  The reconstructed ^{3}H plots for CFC-, SF$_6$- and
  ^{3}H/^{3}He-based ages are shown in figures B105, B106, and
  B107. Most of the samples indicate a mixed groundwater age.
  The ^{3}H/^{3}He-based reconstruction shows evidence of capturing
  the period of time around the bomb peak, with samples
  affected by dispersion.

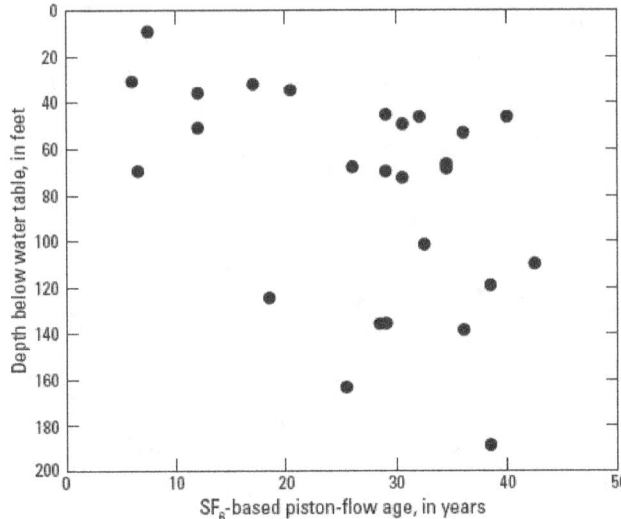

Figure B103.   SF$_6$-based age gradient for dated sites from
the SUS1 network, SACR Study Unit.

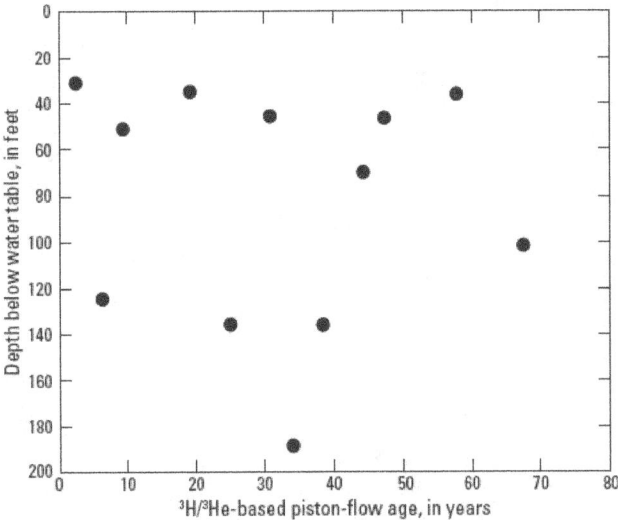

Figure B104. ^{3}H/^{3}He-based age gradient for dated sites
from the SUS1 network, SACR Study Unit.

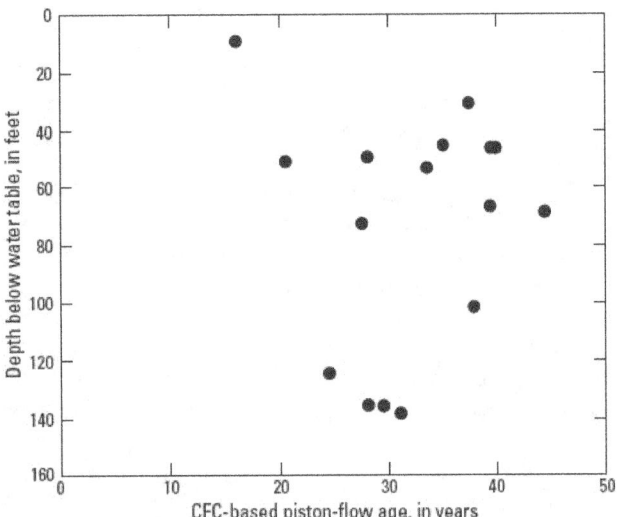

Figure B102.   CFC-based age gradient for dated sites from
the SUS1 network, SACR Study Unit.

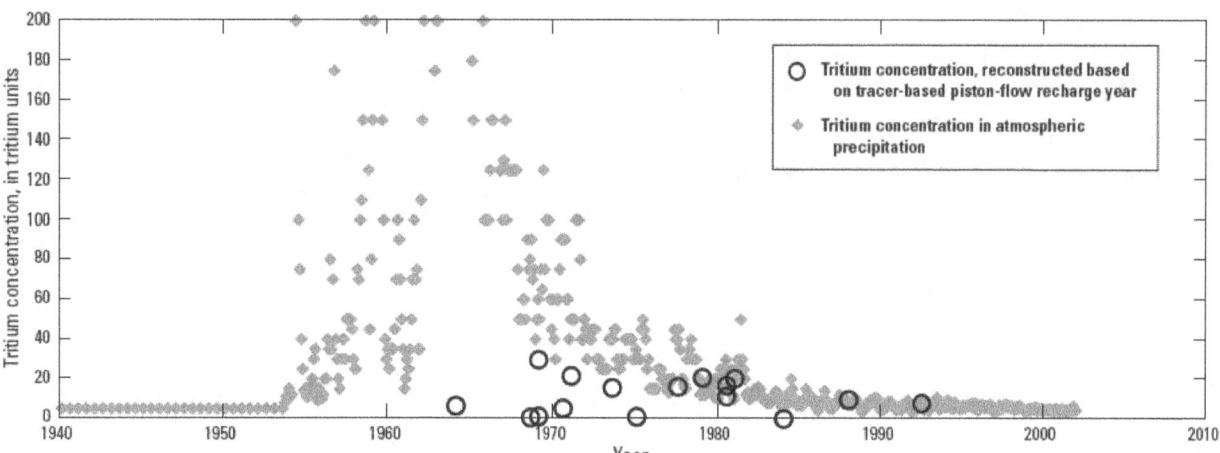

Figure B105. Reconstructed tritium concentrations (using CFC-based ages) and tritium in atmospheric precipitation, SUS1 network, SACR Study Unit.

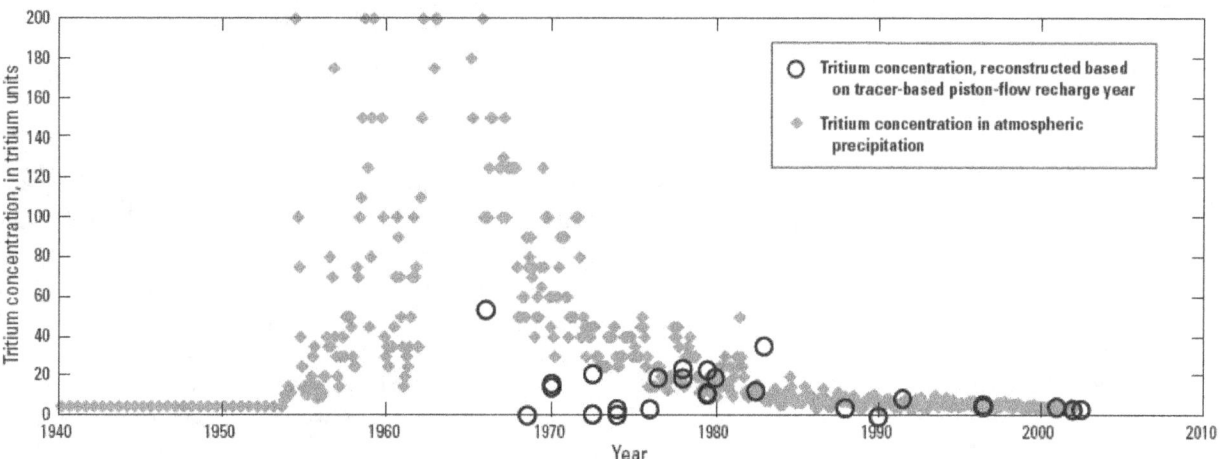

Figure B106. Reconstructed tritium concentrations (using SF_6-based ages) and tritium in atmospheric precipitation, SUS1 network, SACR Study Unit.

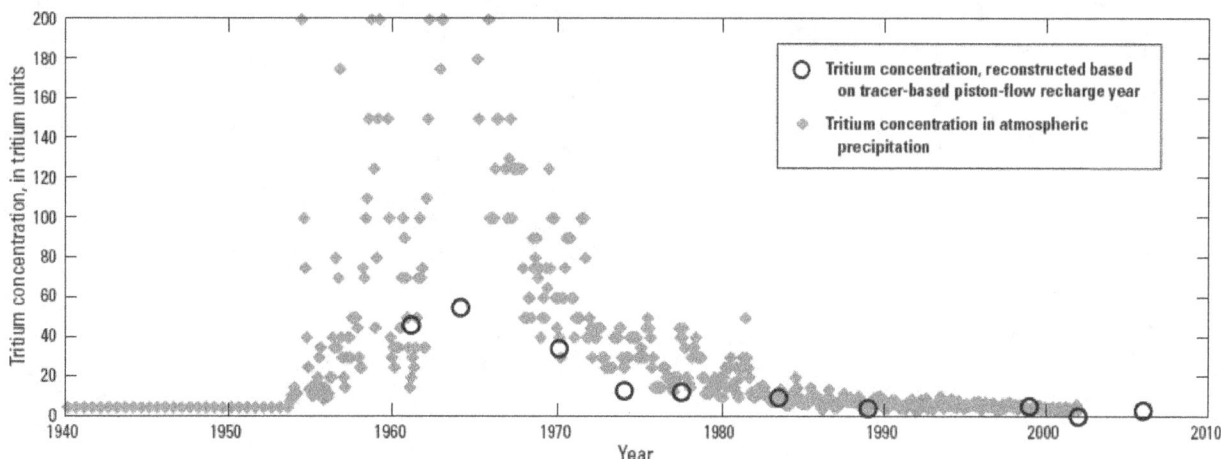

Figure B107. Reconstructed tritium concentrations (using $^3H/^3He$-based ages) and tritium in atmospheric precipitation, SUS1 network, SACR Study Unit.

The SF_6- versus CFC-based age comparison, the $^3H/^3He$-versus CFC-based age comparison, and the $^3H/^3He$-versus SF_6-based age comparison for this network are shown in figures B108, B109, and B110. The age comparison between CFC- and SF_6-based ages is reasonably good, while the comparison between $^3H/^3He$- and CFC-based ages is poor. The comparison between $^3H/^3He$- and SF_6-based ages shows good agreement to the point where the limits of SF_6-based dating is useful, at which point the $^3H/^3He$-based ages continue to get older while the SF_6-based ages do not.

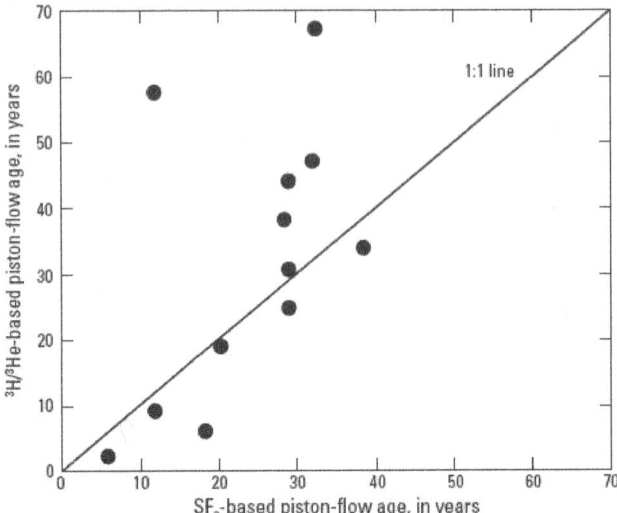

Figure B110. $^3H/^3He$- versus SF_6-based age comparison, SUS1 network, SACR Study Unit.

SANJ LUSCR1a

Samples from five sites in the SANJ Study Unit were collected in 2008 for CFCs, SF_6, and $^3H/^3He$ (networks and, in parentheses, number of sites):

.   LUSCR1a (CFCs and SF_6, 5; $^3H/^3He$, 4)

The aquifer is composed of alluvial sand, gravel, silt and clay of the Central Valley aquifer system.

Major dissolved-gas data were available for all five sites. Of these five sites, four were oxic and one was suboxic.

Age interpretations from tracer concentrations were made assuming that recharge elevation was equal to the elevation of the water table. Estimates of recharge temperature and excess air were based on major dissolved-gas data, with recharge temperature and excess air at suboxic sites being constrained using median excess air at oxic sites.

$^3H/^3He$ ages were calculated for four sites (all four sites did not require a correction for terrigenic He).

The raw tracer data, major dissolved-gas data, the ancillary chemical and well construction data that were used in the interpretations, and the piston-flow ages are presented in table B35.

.   Advantages associated with these samples:

.   Multiple tracers (CFCs, SF_6, and $^3H/^3He$, as well as major dissolved gases).

.   Domestic wells so likely low pumping stress.

.   Disadvantages associated with these samples:

.   Relatively large open intervals ranging from 20 to 40 feet so mixing likely.

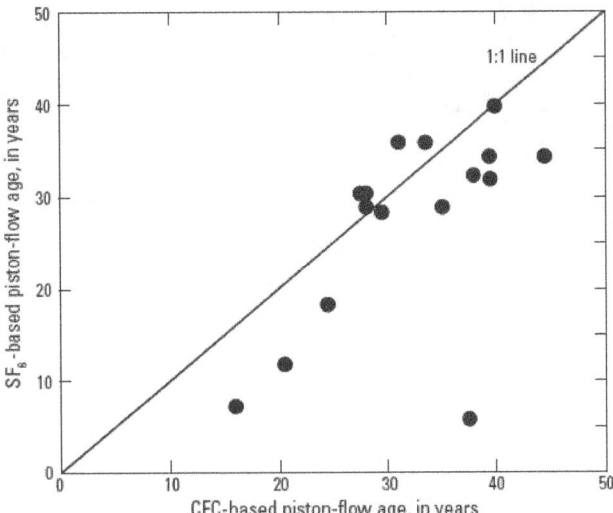

Figure B108. SF_6- versus CFC-based age comparison, SUS1 network, SACR Study Unit.

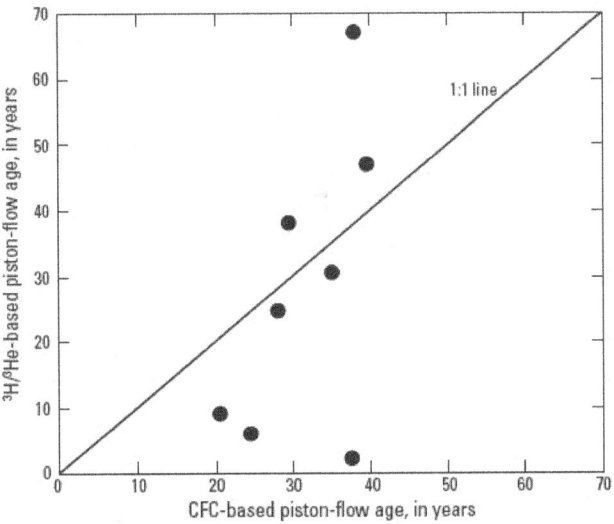

Figure B109. $^3H/^3He$- versus CFC-based age comparison, SUS1 network, SACR Study Unit.

Median penetration of center of open interval into water table was 58.52 feet (not sampling close to the water table, potentially mixing).

Depth to water (can affect tracer transport to water table):

Median: 82.42 feet

Mean: 64.97 feet

Min: 6.80 feet

Max: 111.87 feet

Brief analysis:

The CFC-, SF_6-, and $^3H/^3He$-based age gradients for these sites are shown in figures B111, B112, and B113. The age gradients show a great deal of scatter as would be expected for samples taken from wells with large open intervals and a relatively deep unsaturated zone. Differences in screen length, recharge source/strength, aquifer heterogeneity, pumping stresses, and the position of the well within the flow system may cause some wells to deviate from the general pattern of increasing age with depth. The $^3H/^3He$-based ages are significantly younger as would be expected since helium would be lost in the unsaturated zone until recharge occurs.

The reconstructed ^3H plots for CFC-, SF_6-, and $^3H/^3He$-based ages are shown in figures B114, B115, and B116. The reconstructions show a similar pattern to the age gradients above in that the CFC- and SF_6-based ages

are similar, and the $^3H/^3He$-based ages are very young. The samples generally plot along the tritium reconstructions, with minor offsets that likely result from sampling large open intervals with deep unsaturated zones.

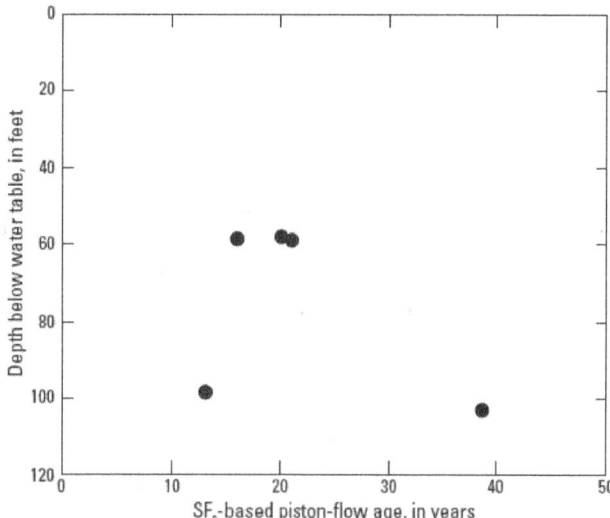

Figure B112. SF_6-based age gradient for dated sites from the LUSCR1a network, SANJ Study Unit.

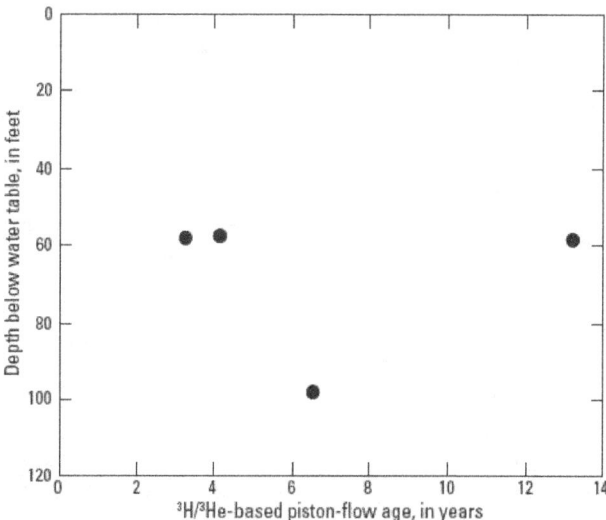

Figure B113. $^3H/^3He$-based age gradient for dated sites from the LUSCR1a network, SANJ Study Unit.

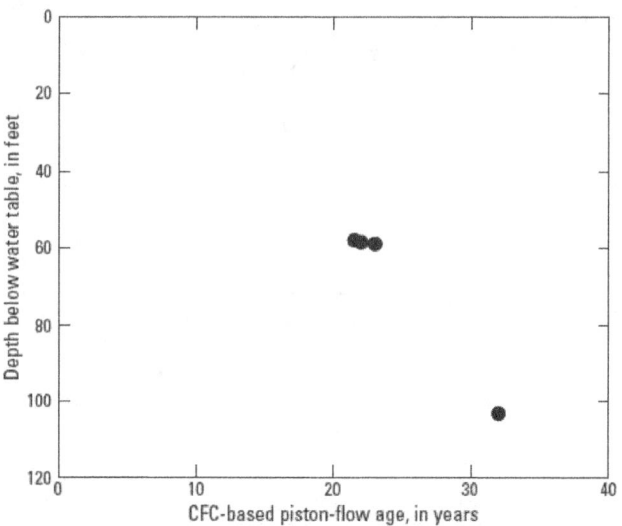

Figure B111. CFC-based age gradient for dated sites from the LUSCR1a network, SANJ Study Unit.

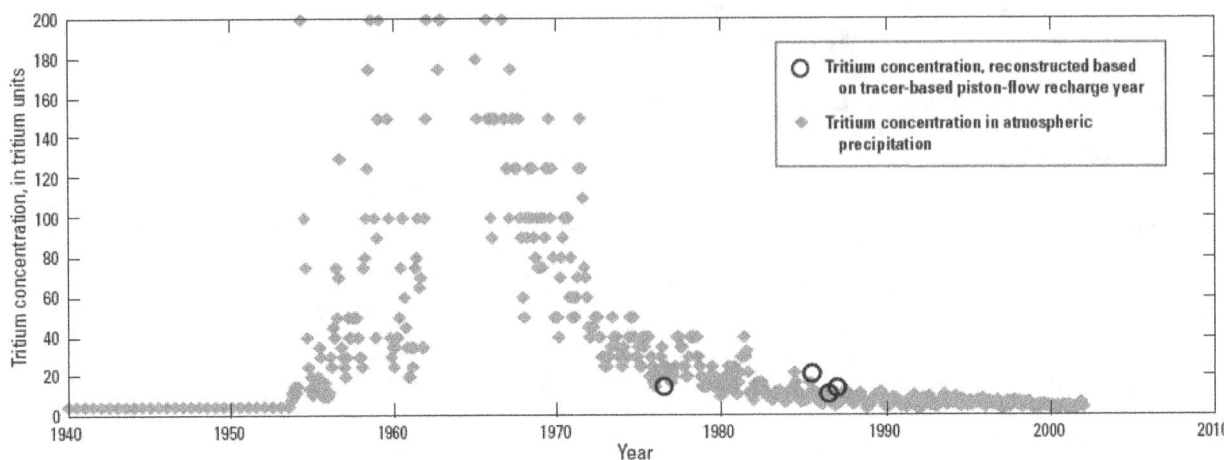

Figure B114.   Reconstructed tritium concentrations (using CFC-based ages) and tritium in atmospheric precipitation, LUSCR1a network, SANJ Study Unit.

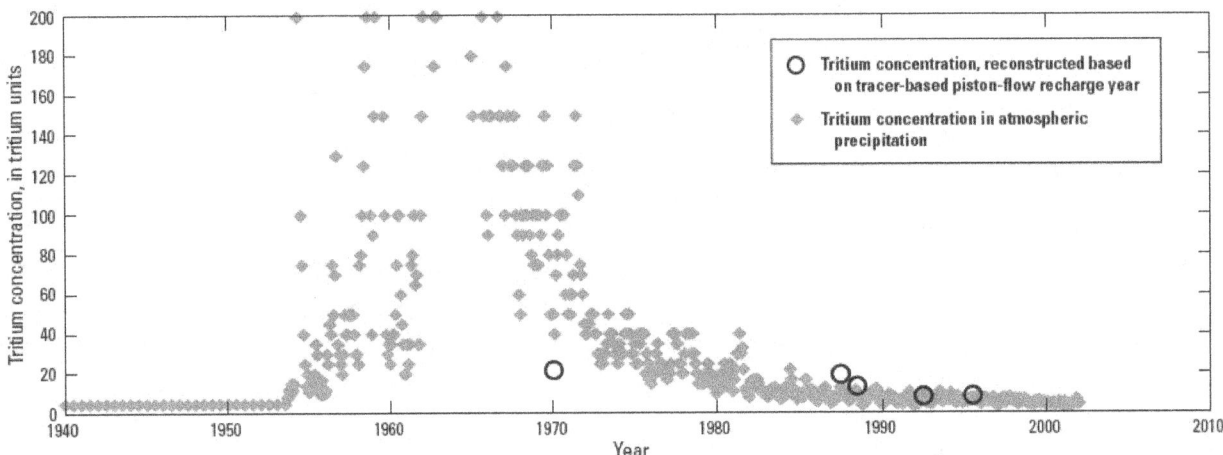

Figure B115.   Reconstructed tritium concentrations (using SF$_6$-based ages) and tritium in atmospheric precipitation, LUSCR1a network, SANJ Study Unit.

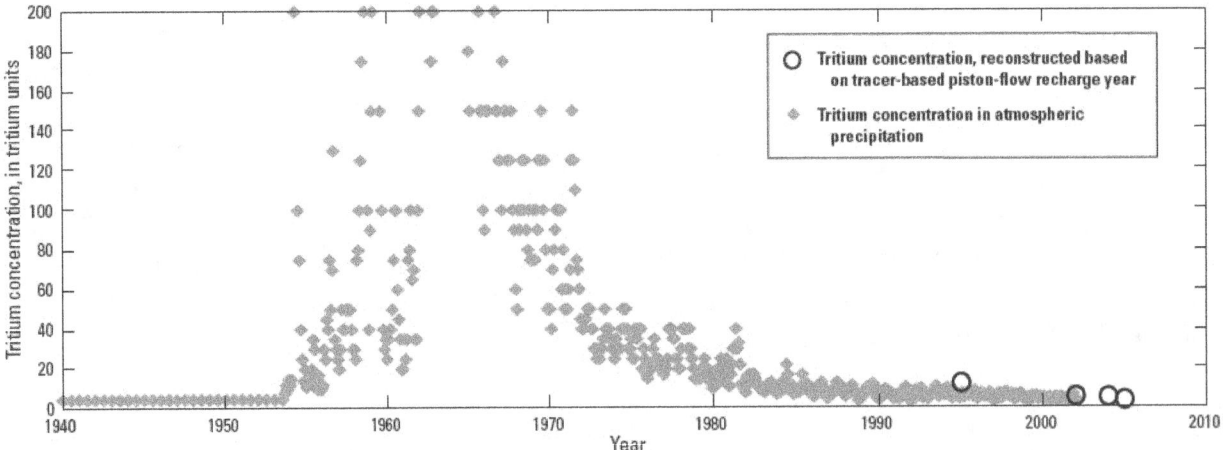

Figure B116.   Reconstructed tritium concentrations (using ^{3}H/^{3}He-based ages) and tritium in atmospheric precipitation, LUSCR1a network, SANJ Study Unit.

The SF$_6$- versus CFC-based age comparison, the ^{3}H/^{3}He- versus CFC-based age comparison, and the ^{3}H/^{3}He- versus SF$_6$-based age comparison for this network are shown in figures B117, B118, and B119. The age comparisons show reasonable agreement for CFC- and SF$_6$-based ages, and a bias toward younger ages for ^{3}H/^{3}He-based ages as would be expected for samples taken from wells with a deep unsaturated zone.

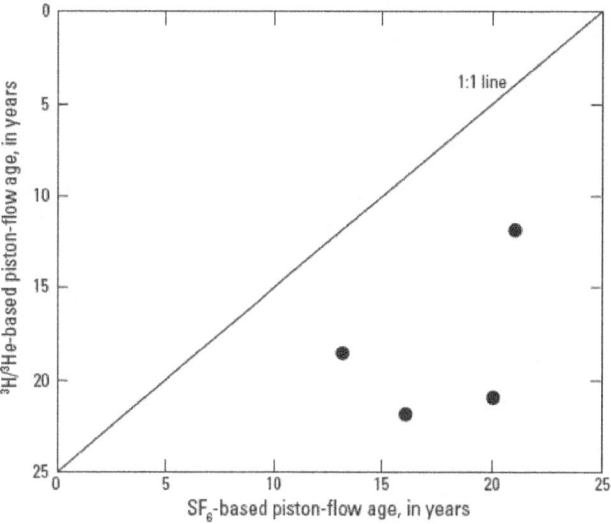

Figure B119. ^{3}H/^{3}He- versus SF$_6$-based age comparison, LUSCR1a network, SANJ Study Unit.

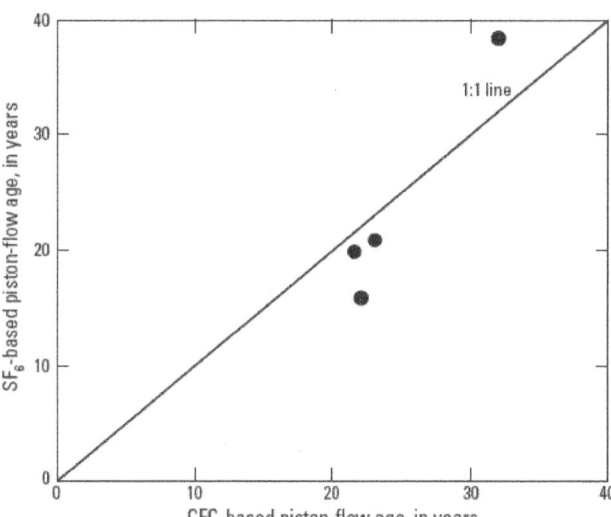

Figure B117.   SF$_6$- versus CFC-based age comparison, LUSCR1a network, SANJ Study Unit.

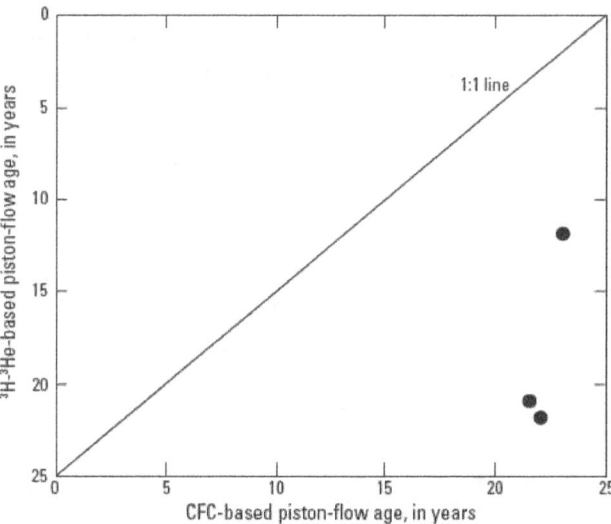

Figure B118. ^{3}H/^{3}He- versus CFC-based age comparison, LUSCR1a network, SANJ Study Unit.

SANJ LUSOR1a

Samples from five sites in the SANJ Study Unit were collected in 2006 for CFCs and SF$_6$, and from one site for ^{3}H/^{3}He in 2008 (networks and, in parentheses, number of sites):

. LUSOR1a (CFCs and SF$_6$, 5 in 2006; ^{3}H/^{3}He, 1 in 2008)

The aquifer is composed of alluvial sand, gravel, silt and clay of the Central Valley aquifer system.

Major dissolved-gas data were available for all five sites. Of these five sites, all five were oxic.

Age interpretations from tracer concentrations were made assuming that recharge elevation was equal to the elevation of the water table. Estimates of recharge temperature and excess air were based on major dissolved-gas data.

^{3}H/^{3}He age was calculated for only one site and it did not require a correction for terrigenic He.

The raw tracer data, major dissolved-gas data, the ancillary chemical and well construction data that were used in the interpretations, and the piston-flow ages are presented in table B36.

. Advantages associated with these samples:

. Multiple tracers (CFCs, SF$_6$, and ^{3}H/^{3}He, as well as major dissolved gases).

. Domestic wells so likely low pumping stress.

. Disadvantages associated with these samples:

. Relatively large open intervals ranging from 14 to 50 feet so mixing likely.

.   Median penetration of center of open interval into water table was 74.08 feet (not sampling close to the water table, potentially mixing).

.   Depth to water (can affect tracer transport to water table):

.   Median: 43.73 feet

.   Mean: 43.29 feet

.   Min: 34.64 feet

.   Max: 51.87 feet

.   Brief analysis:

.   The CFC- and SF$_6$-based age gradients for these sites are shown in figures B120 and B121. The age

gradients show little structure, but are similar for both CFC- and SF$_6$-based dating. Differences in screen length, recharge source/strength, aquifer heterogeneity, pumping stresses, and the position of the well within the flow system may cause some wells to deviate from the general pattern of increasing age with depth.

The reconstructed ^{3}H plots for CFC- and SF$_6$-based ages are shown in figures B122 and B123. The reconstructions are similar and show evidence of relatively unmixed, piston-flow transport.

The SF$_6$- versus CFC-based age comparison for this network is shown in figure B124. The age comparison shows consistency between CFC- and SF$_6$-based ages.

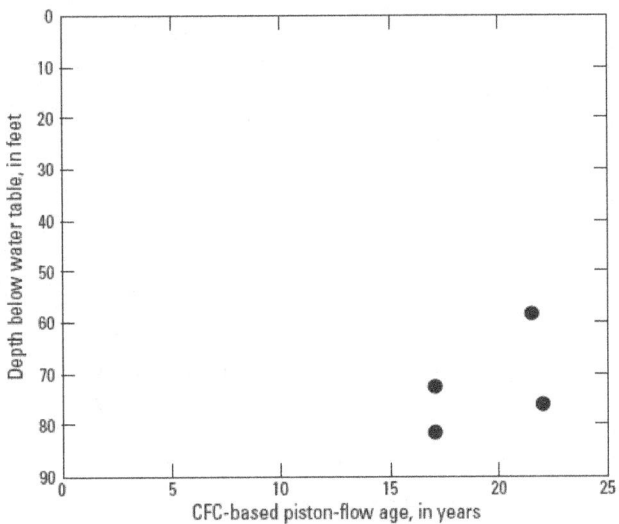

Figure B120.   CFC-based age gradient for dated sites from the LUSOR1a network. SANJ Study Unit.

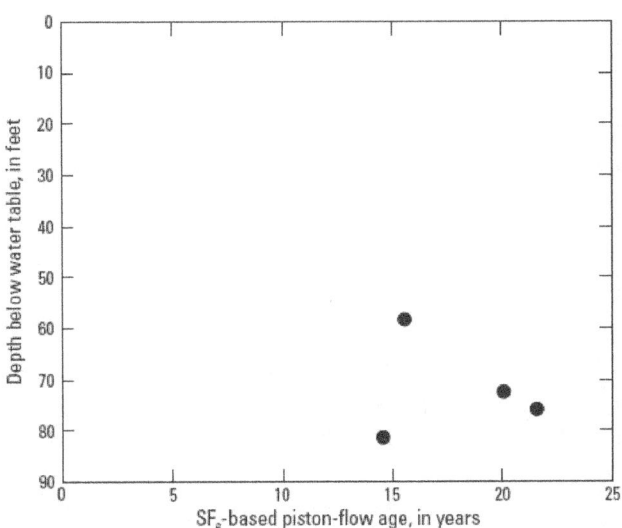

Figure B121.   SF$_6$-based age gradient for dated sites from the LUSOR1a network, SANJ Study Unit.

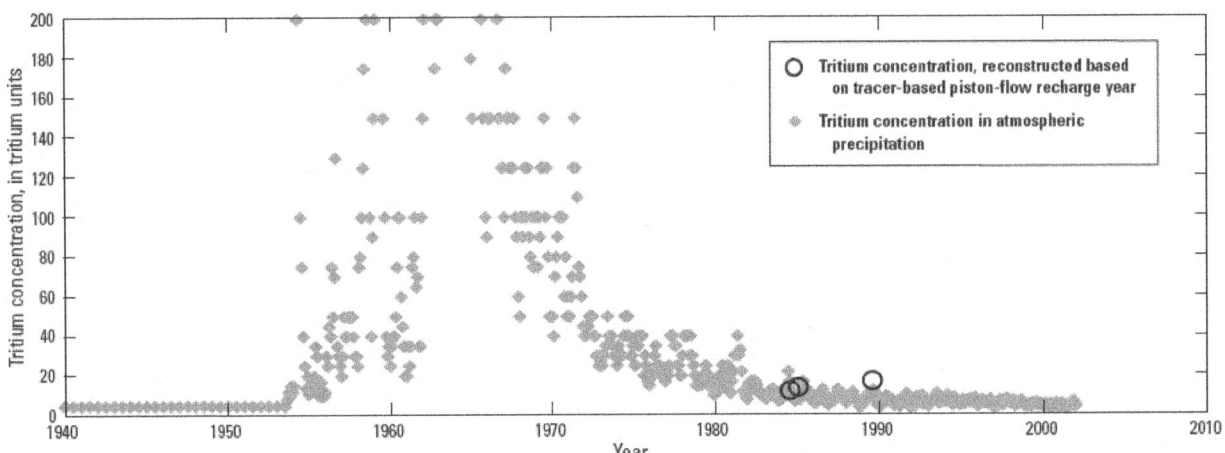

Figure B122.   Reconstructed tritium concentrations (using CFC-based ages) and tritium in atmospheric precipitation, LUSOR1a network, SANJ Study Unit.

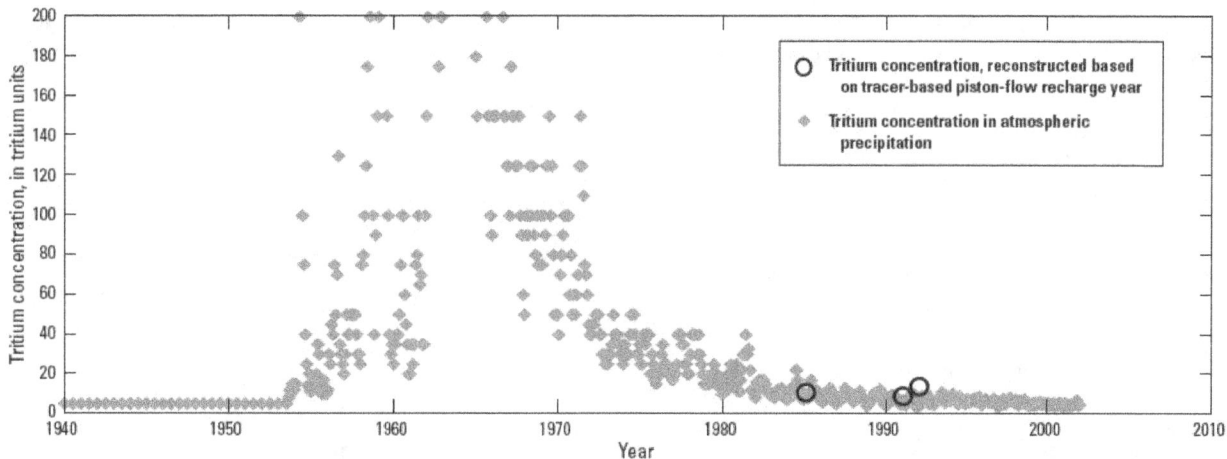

Figure B123. Reconstructed tritium concentrations (using SF_6-based ages) and tritium in atmospheric precipitation, LUSOR1a network, SANJ Study Unit.

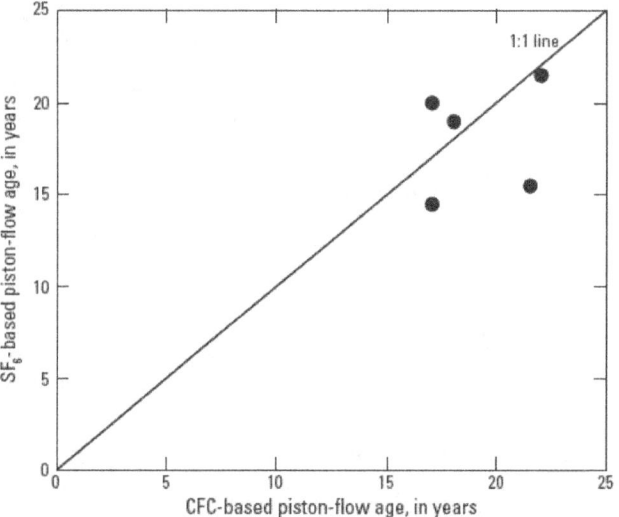

Figure B124. SF_6- versus CFC-based age comparison, LUSOR1a network, SANJ Study Unit.

SANJ LUSOR2a

Samples from five sites in the SANJ Study Unit were collected in 2006 for CFCs and SF_6, and from two sites in 2008 for $^3H/^3He$ (networks and, in parentheses, number of sites):

. LUSOR2a (CFCs and SF_6, 5; $^3H/^3He$, 2)

The aquifer is composed of alluvial sand, gravel, silt, and clay of the Central Valley aquifer system.

Major dissolved-gas data were available for all five sites in 2006 and two sites in 2008. Of these five sites, four were oxic and one was suboxic in 2006, and both sites were oxic in 2008.

Age interpretations from tracer concentrations were made assuming that recharge elevation was equal to the elevation of the water table. Estimates of recharge temperature and excess air were based on major dissolved-gas data, with recharge temperature and excess air at suboxic sites being constrained using median excess air at oxic sites.

$^3H/^3He$ ages were calculated for two sites (only one site required a correction for terrigenic helium).

The raw tracer data, major dissolved-gas data, the ancillary chemical and well construction data that were used in the interpretations, and the piston-flow ages are presented in table B37.

. Advantages associated with these samples:

. Multiple tracers (CFCs, SF_6, and $^3H/^3He$, as well as major dissolved gases).

. Domestic wells so likely low pumping stress.

. Disadvantages associated with these samples:

. Open intervals ranging from 10 to 20 feet so some mixing likely.

. Median penetration of center of open interval into water table was 138.89 feet (not sampling close to the water table, potentially mixing).

. Depth to water (can affect tracer transport to water table):

. Median: 61.00 feet

. Mean: 53.13 feet

. Min: 8.46 feet

. Max: 82.93 feet

. Brief analysis:

. The SF_6-based age gradient for these sites is shown in figure B125. The age gradient shows no real structure. Differences in screen length, recharge source/strength, aquifer heterogeneity, pumping stresses, and the position of the well within the flow system may cause some wells to deviate from the general pattern of increasing age with depth. The CFC- and $^3H/^3He$-based age gradients are not shown because there were only two samples to plot in each case.

The reconstructed ^3H plots for CFC-, SF_6- and $^3H/^3He$-based ages are shown in figures B126, B127, and B128. Numerous samples plot above the ^3H reconstruction as might be expected for samples taken from sites with large open intervals and relatively deep unsaturated zones. The $^3H/^3He$-based reconstruction, though only two samples, is excellent.

The SF_6- versus CFC-based age comparison for this network is shown in figure B129. The age comparison is limited due to the low number of samples, but is consistent between CFC- and SF_6-based ages.

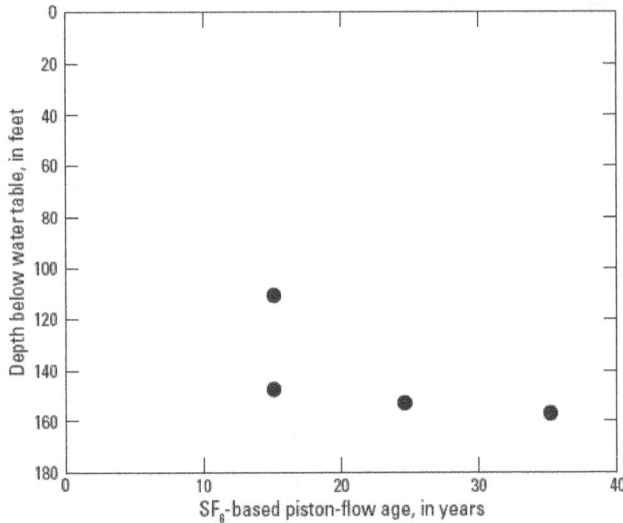

Figure B125. SF_6-based age gradient for dated sites from the LUSOR2a network, SANJ Study Unit.

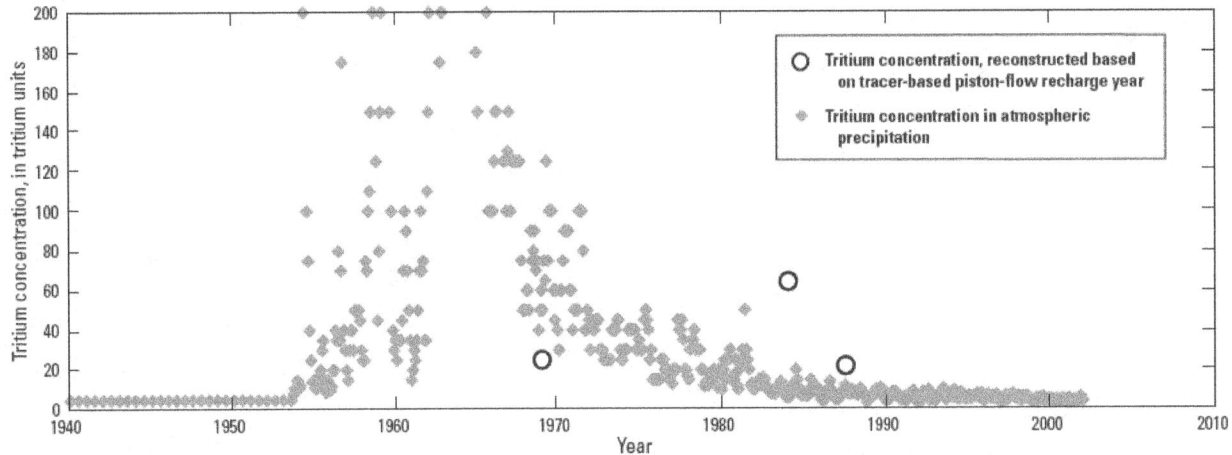

Figure B126. Reconstructed tritium concentrations (using CFC-based ages) and tritium in atmospheric precipitation, LUSOR2a network, SANJ Study Unit.

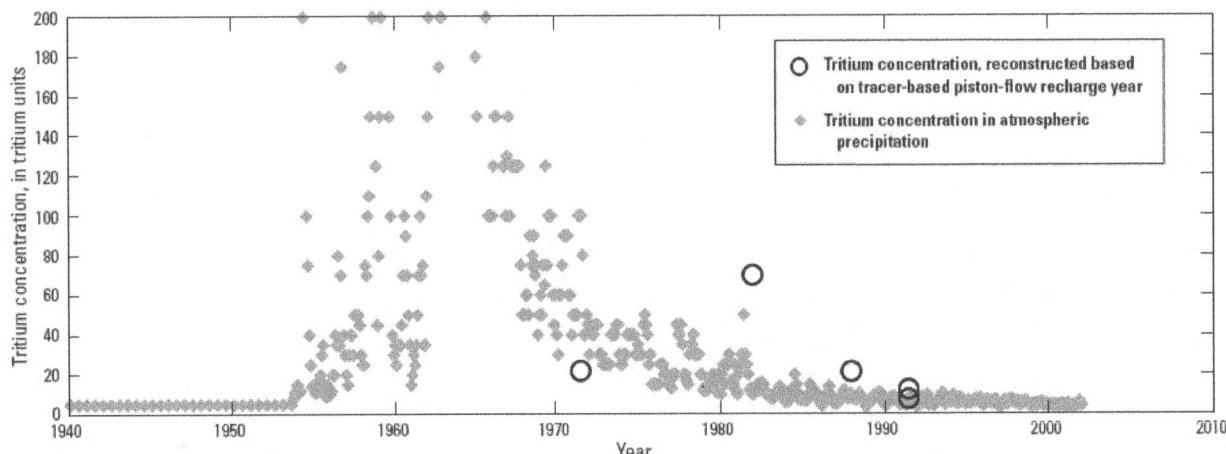

Figure B127. Reconstructed tritium concentrations (using SF_6-based ages) and tritium in atmospheric precipitation, LUSOR2a network, SANJ Study Unit.

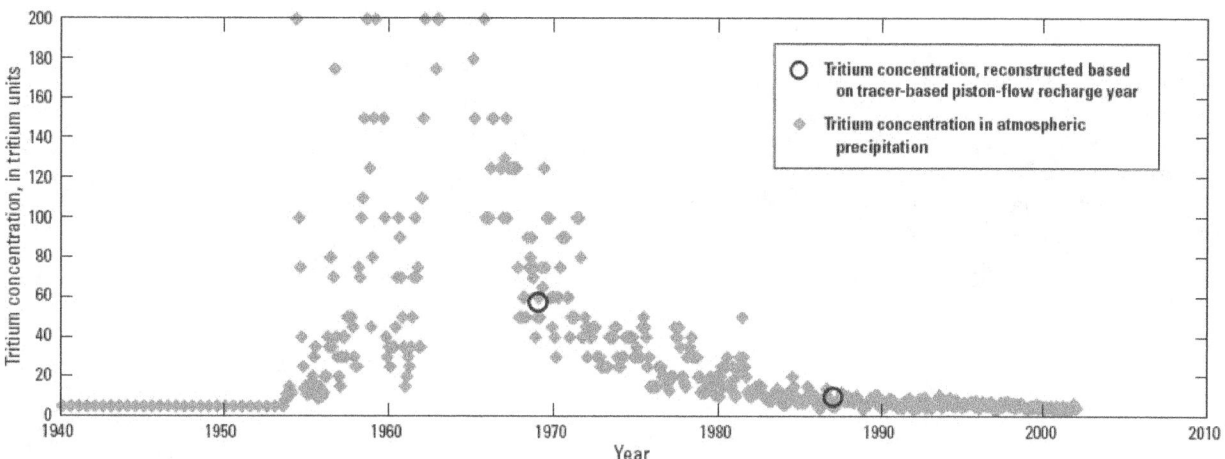

Figure B128. Reconstructed tritium concentrations (using $^3H/^3He$-based ages) and tritium in atmospheric precipitation, LUSOR2a network, SANJ Study Unit.

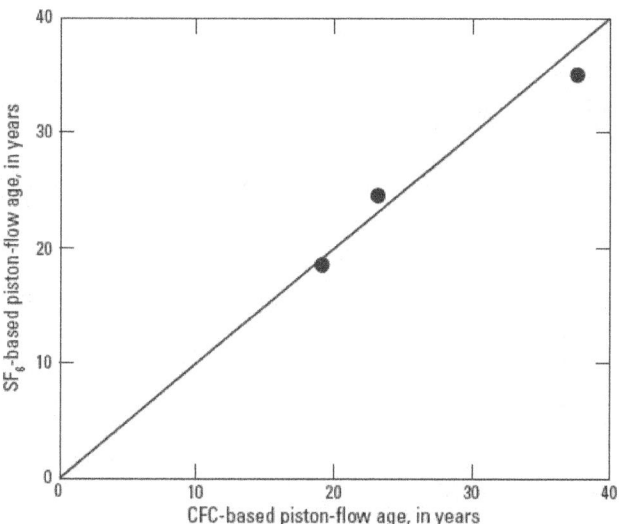

Figure B129. SF$_6$- versus CFC-based age comparison, LUSOR2a network, SANJ Study Unit.

SANJ SUS1

Samples from five sites in the SANJ Study Unit were collected in 2008 for CFCs, SF$_6$, and $^3H/^3He$ (networks and, in parentheses, number of sites):

. SUS1 (CFCs and SF$_6$, 5; $^3H/^3He$, 4)

The aquifer is composed of alluvial sand, gravel, silt and clay of the Central Valley aquifer system.

Major dissolved-gas data were available for all five sites. Of these five sites, all five were oxic.

Age interpretations from tracer concentrations were made assuming that recharge elevation was equal to the elevation of the water table. Estimates of recharge temperature and excess air were based on major dissolved-gas data.

$^3H/^3He$ ages were calculated for four sites (all four sites did not require a correction for terrigenic He).

The raw tracer data, major dissolved-gas data, the ancillary chemical and well construction data that were used in the interpretations, and the piston-flow ages are presented in table B38.

. Advantages associated with these samples:

. Multiple tracers (CFCs, SF^6, and $^3H/^3He$, as well as major dissolved gases).

. Domestic wells so likely low pumping stress.

. Disadvantages associated with these samples:

. Relatively large open intervals ranging from 18 to 100 feet so mixing likely.

. Median penetration of center of open interval into water table was 66.87 feet (not sampling close to the water table, potentially mixing).

. Depth to water (can affect tracer transport to water table):

. Median: 63.60 feet

. Mean: 88.74 feet

. Min: 34.13 feet

. Max: 145.34 feet

. Brief analysis:

. The CFC-, SF$_6$-, and $^3H/^3He$-based age gradients for these sites are shown in figures B130, B131, and B132. The age gradients show a general increase in age with depth. Differences in screen length, recharge source/strength, aquifer heterogeneity, pumping stresses, and the position of the well within the flow system may cause some wells to deviate from the general pattern of increasing age with depth.

The reconstructed ^3H plots for CFC-, SF_6-, and $^3H/^3He$-based ages are shown in figures B133, B134, and B135. The reconstructions show evidence of relatively unmixed, piston-flow transport.

The SF_6- versus CFC-based age comparison, the $^3H/^3He$- versus CFC-based age comparison, and the $^3H/^3He$- versus SF_6-based age comparison for this network are shown in figures B136, B137, and B138. All three age comparisons show a general consistency, but some dispersive effects also are apparent.

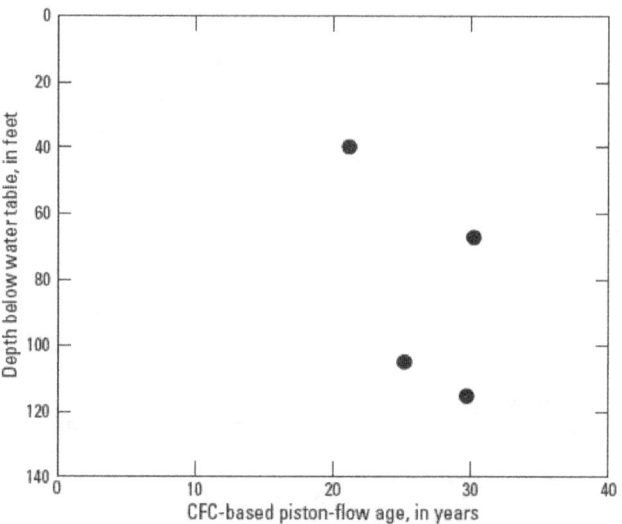

Figure B130. CFC-based age gradient for dated sites from the SUS1 network, SANJ Study Unit.

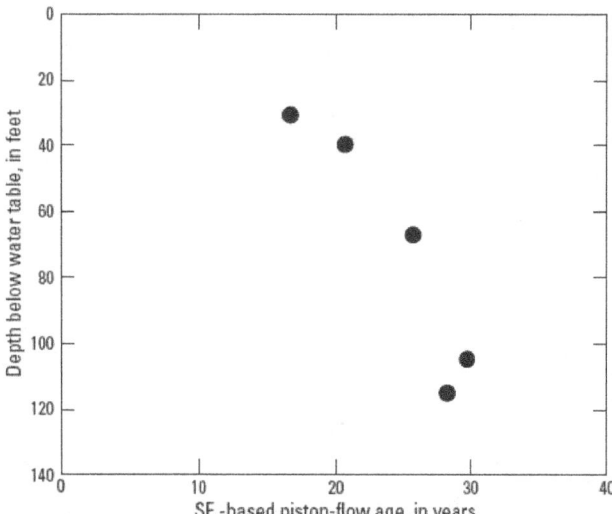

Figure B131. SF_6-based age gradient for dated sites from the SUS1 network, SANJ Study Unit.

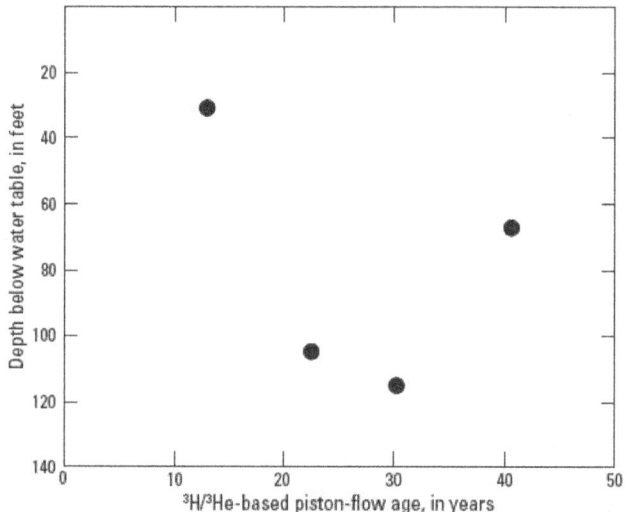

Figure B132. $^3H/^3He$-based age gradient for dated sites from the SUS1 network, SANJ Study Unit.

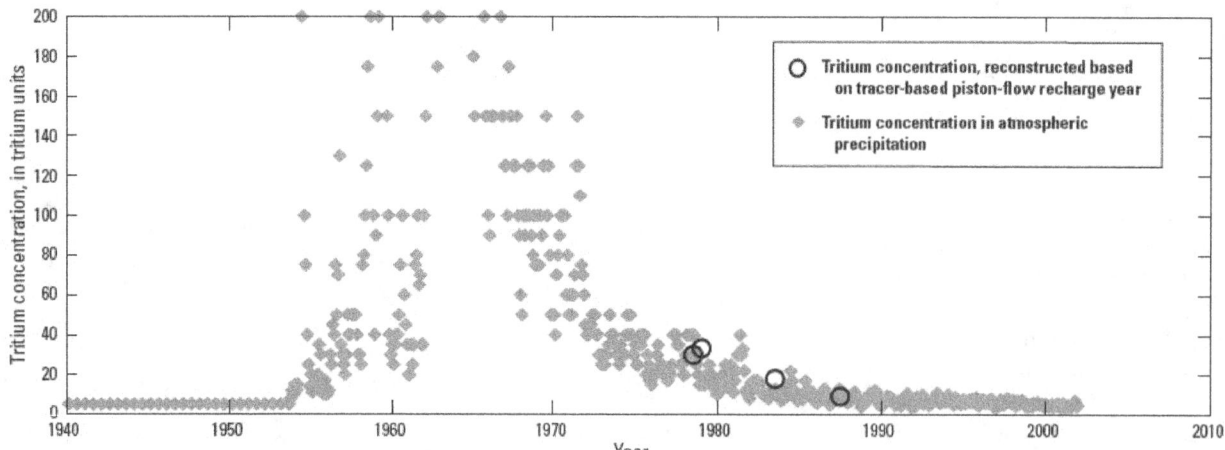

Figure B133. Reconstructed tritium concentrations (using CFC-based ages) and tritium in atmospheric precipitation, SUS1 network, SANJ Study Unit.

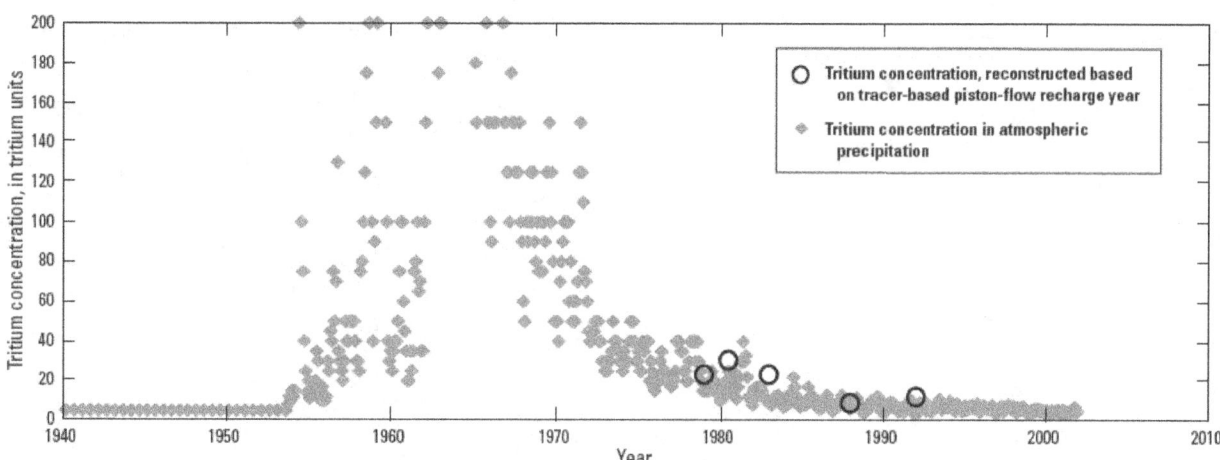

Figure B134. Reconstructed tritium concentrations (using SF$_6$-based ages) and tritium in atmospheric precipitation, SUS1 network, SANJ Study Unit.

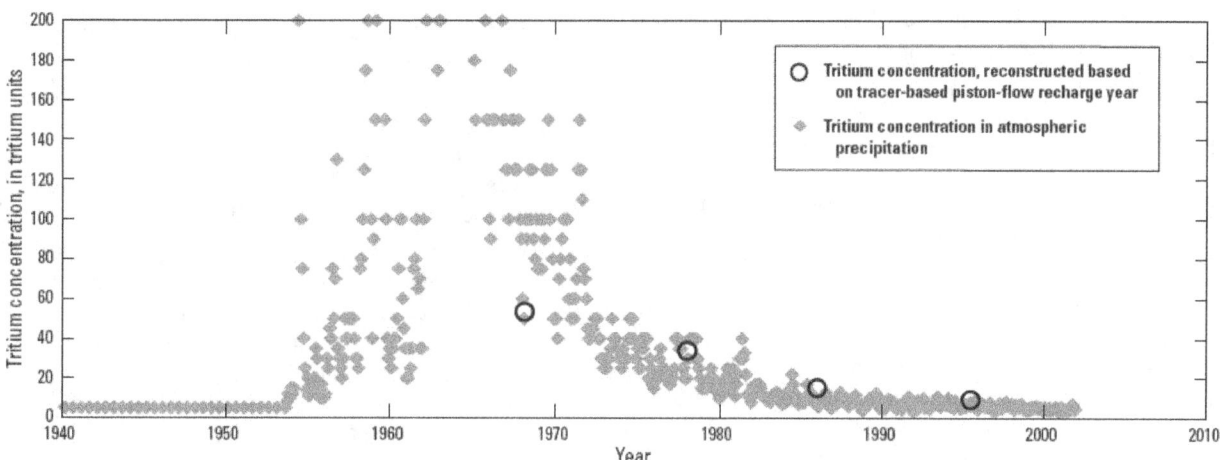

Figure B135. Reconstructed tritium concentrations (using ^{3}H/^{3}He-based ages) and tritium in atmospheric precipitation, SUS1 network, SANJ Study Unit.

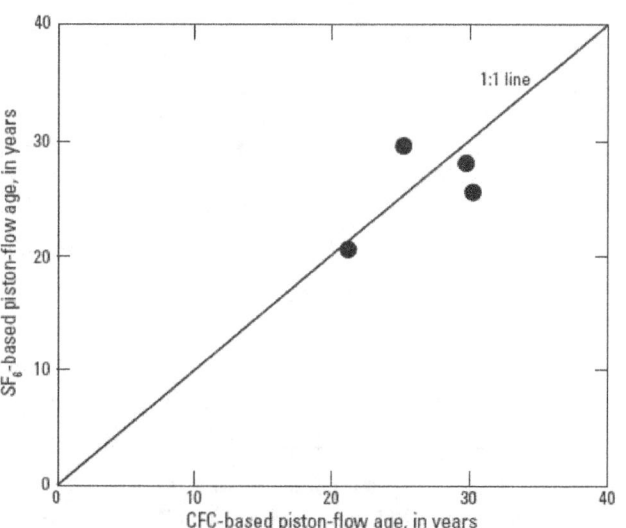

Figure B136. SF$_6$- versus CFC-based age comparison, SUS1 network, SANJ Study Unit.

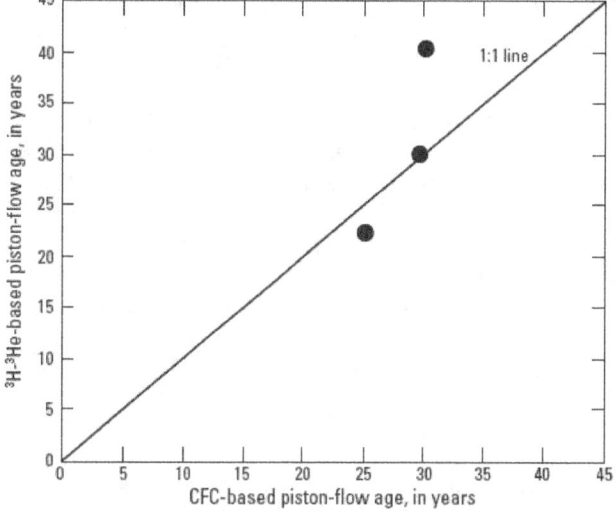

Figure B137. ^{3}H/^{3}He- versus CFC-based age comparison, SUS1 network, SANJ Study Unit.

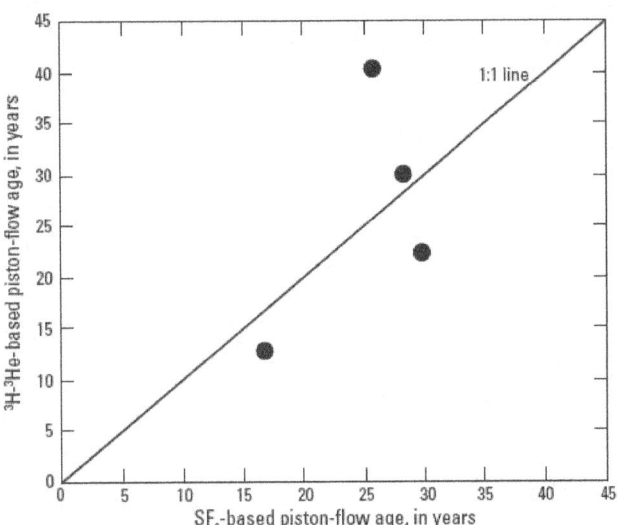

Figure B138.   ³H/³He- versus SF₆-based age comparison, SUS1 network, SANJ Study Unit.

SANT FPSAG1

Samples from 16 sites in the SANT Study Unit were collected in 2008 for CFCs and SF₆ (networks and, in parentheses, number of sites):

.   FPSAG1 (CFCs, 16; SF₆, 1)

The aquifer is composed of sands, and clay with some sand. Two sites were finished in limestone.

Major dissolved-gas data were available for 16 sites. Of these 16 sites, 9 were oxic and 7 were suboxic.

Age interpretations from tracer concentrations were made assuming that recharge elevation was equal to the elevation of the water table. Estimates of recharge temperature and excess air were based on major dissolved-gas data, with recharge temperature and excess air at suboxic sites being constrained using median excess air at oxic sites. Using this approach, however, many of the sites with elevated methane concentrations had questionable dissolved-gas results. A comparison was done between results using MAAT+1°C and major dissolved-gas data, with similar results using both methods, but slightly older results using the MAAT+1°C for two sites.

The raw tracer data, major dissolved-gas data, the ancillary chemical and well construction data that were used in the interpretations, and the piston-flow ages are presented in table B39.

.   Advantages associated with these samples:

.   Multiple tracers (CFCs and SF₆, as well as major dissolved gases).

.   Monitoring wells, therefore low pumping stress.

.   Short open intervals of 0.5 foot so mixing likely minimized.

.   Median penetration of center of open interval into water table was 8.02 feet (sampling close to the water table, potentially minimizes mixing).

.   Disadvantages associated with these samples:

.   Suboxic conditions.

.   Depth to water (can affect tracer transport to water table):

.   Median: 6.96 feet

.   Mean: 5.54 feet

.   Min: -0.29 feet

.   Max: 10.23 feet

.   Brief analysis:

The CFC-based age gradient for these sites is shown in figure B139. The age gradient shows a general increase in age with depth with the exception of two samples, both of which are located in surface-water bodies (stream channel and lake/swamp) indicating that they may be in discharge areas and would be expected to have older ages at shallow depths. The CFC-based ages also appear to be shifted toward older ages, which may be the result of suboxic conditions in numerous wells.

The reconstructed ³H plot for CFC-based ages is shown in figure B140. Numerous samples plot above and/or to the right of the ³H reconstruction, which may be the result of CFC-based ages being too old as a result of degradation. If the CFC-based ages were younger, the samples would shift down and to the right, however, the reconstructed tritium would still be somewhat elevated. The reason for the slightly elevated tritium in these samples is unknown.

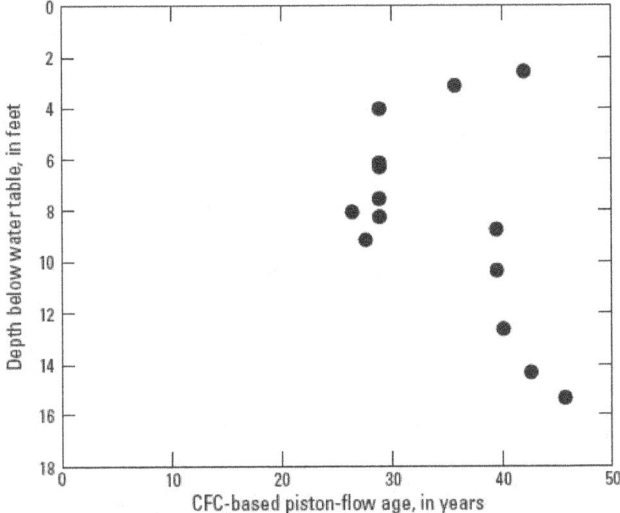

Figure B139.   CFC-based age gradient for dated sites from the FPSAG1 network, SANT Study Unit.

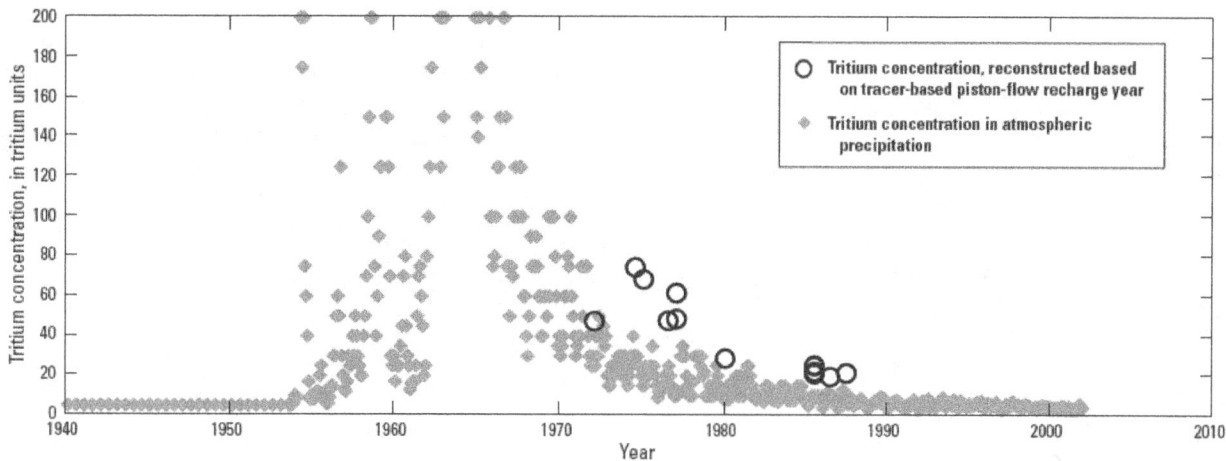

Figure B140. Reconstructed tritium concentrations (using CFC-based ages) and tritium in atmospheric precipitation, FPSAG1 network, SANT Study Unit.

SANT LUSRC1

Samples from 19 sites in the SANT Study Unit were collected in 2006 for CFCs and SF_6 (networks and, in parentheses, number of sites):

. LUSRC1 (CFCs, 19; SF_6, 4)

The aquifer is composed of clay with some sand of the Southeastern Coastal Plain aquifer system.

Major dissolved-gas data were available for 19 sites. Of these 19 sites, 10 were oxic and 9 were suboxic.

Age interpretations from tracer concentrations were made assuming that recharge elevation was equal to the elevation of the water table. Estimates of recharge temperature and excess air were based on major dissolved-gas data, with recharge temperature and excess air at suboxic sites being constrained using median excess air at oxic sites. Using this approach was problematic and the MAAT+1°C, which was similar, was used instead.

The raw tracer data, major dissolved-gas data, the ancillary chemical and well construction data that were used in the interpretations, and the piston-flow ages are presented in table B40.

. Advantages associated with these samples:

. Multiple tracers (CFCs and SF_6, as well as major dissolved gases).

. Monitoring wells, therefore low pumping stress.

. Relatively short open intervals ranging from 2.32 to 5 feet so mixing likely minimized.

. Median penetration of center of open interval into water table was 4.36 feet (sampling close to the water table, potentially minimizing mixing).

. Disadvantages associated with these samples:

. Suboxic conditions.

. No tritium.

. Depth to water (can affect tracer transport to water table):

. Median: 9.70 feet

. Mean: 13.21 feet

. Min: 2.14 feet

. Max: 47.69 feet

. Brief analysis:

. The CFC-based age gradient for these sites is shown in figure B141. The age gradient shows a general increase in age with depth, however, the ages are shifted to older ages even at shallow depths. The relatively old ages at shallow depths may be the result of degradation of the CFCs in suboxic conditions. Differences in screen length, recharge source/strength, aquifer heterogeneity, pumping stresses, and the position of the well within the flow system may cause some wells to deviate from the general pattern of increasing age with depth.

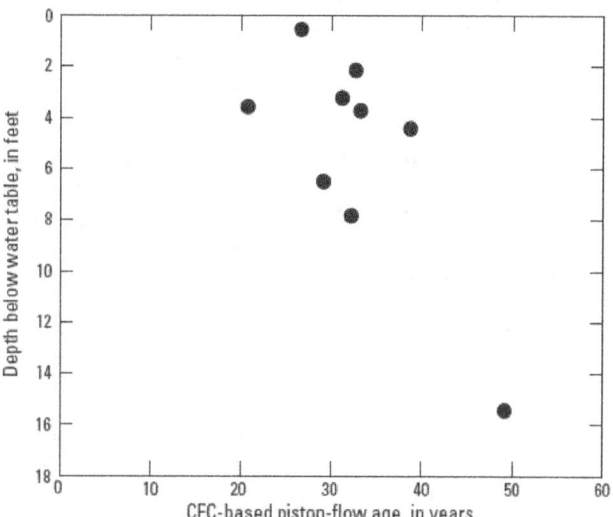

Figure B141. CFC-based age gradient for dated sites from the LUSRC1 network, SANT Study Unit.

SCTX LUSRC1

Samples from 22 sites in the SCTX Study Unit were collected in 2006 for CFCs and SF_6 (networks and, in parentheses, number of sites):

. LUSRC1 (22)

The aquifer is composed of limestone and dolomite of the Edwards-Trinity aquifer system.

Major dissolved-gas data were available for all 22 sites. Of these 22 sites, all 22 were oxic.

Age interpretations from tracer concentrations were made assuming that recharge elevation was equal to the elevation of the water table. Estimates of recharge temperature and excess air were based on major dissolved-gas data. Using this approach, several sites had unreasonably high excess air, but if excess nitrogen was utilized, the recharge temperatures were too low. Also, four sites had unreasonably high recharge temperatures.

The raw tracer data, major dissolved-gas data, the ancillary chemical and well construction data that were used in the interpretations, and the piston-flow ages are presented in table B41.

. Advantages associated with these samples:

. Multiple tracers (CFCs and SF_6, as well as major dissolved gases).

. Monitoring wells, therefore low pumping stress.

. Disadvantages associated with these samples:

. Relatively large open intervals ranging from 34.2 to 84.96 feet so mixing likely.

. Median penetration of center of open interval into water table was 25.92 feet (not sampling close to the water table, potentially mixing).

. Depth to water (can affect tracer transport to water table):

. Median: 221.01 feet

. Mean: 219.56 feet

. Min: 153.09 feet

. Max: 270.25 feet

. Brief analysis:

. The CFC- and SF_6-based age gradients for these sites are shown in figures B142 and B143. The age gradients show a great deal of scatter as would be expected for samples taken from wells finished in karst with large open intervals.

The reconstructed ^3H plots for CFC- and SF_6-based ages are shown in figures B144 and B145. The samples appear to be relatively unmixed, they plot in a region where they could shift to older or younger ages and still plot on the ^3H input function and therefore could still be affected by mixing.

The SF_6- versus CFC-based age comparison for this network is shown in figure B146. The age comparison is very scattered as would be expected for a karst environment where mixing, and perhaps even degassing, is occurring.

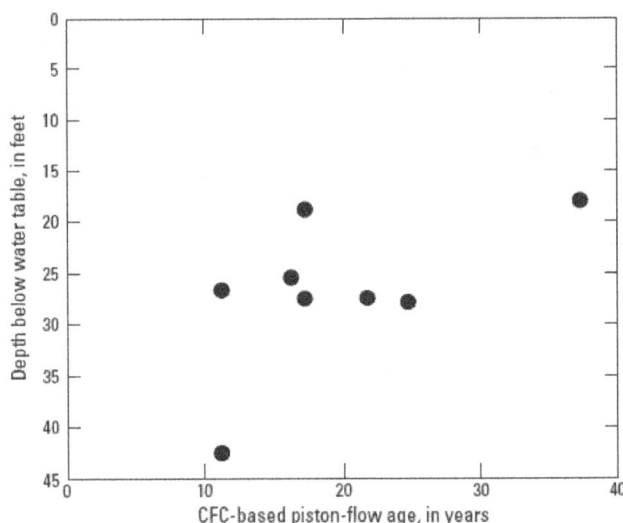

Figure B142. CFC-based age gradient for dated sites from the LUSRC1 network, SCTX Study Unit.

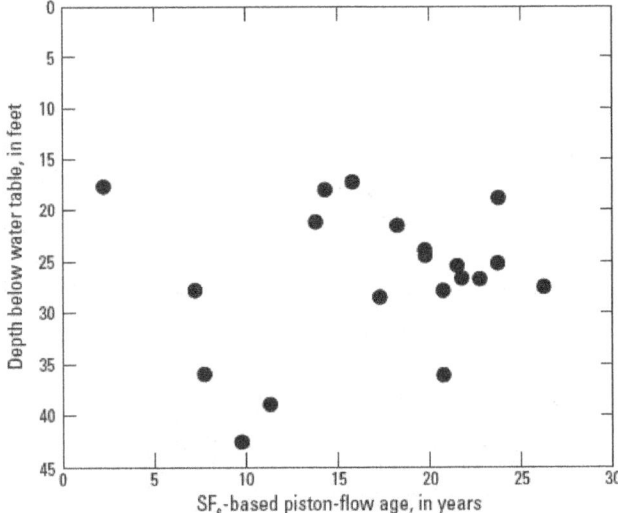

Figure B143. SF_6-based age gradient for dated sites from the LUSRC1 network, SCTX Study Unit.

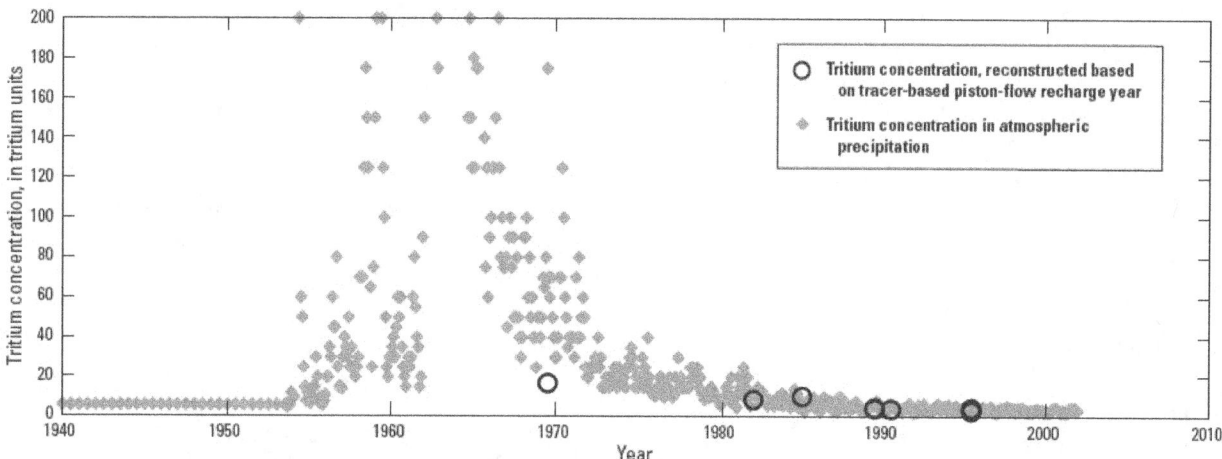

Figure B144.   Reconstructed tritium concentrations (using CFC-based ages) and tritium in atmospheric precipitation, LUSRC1 network, SCTX Study Unit.

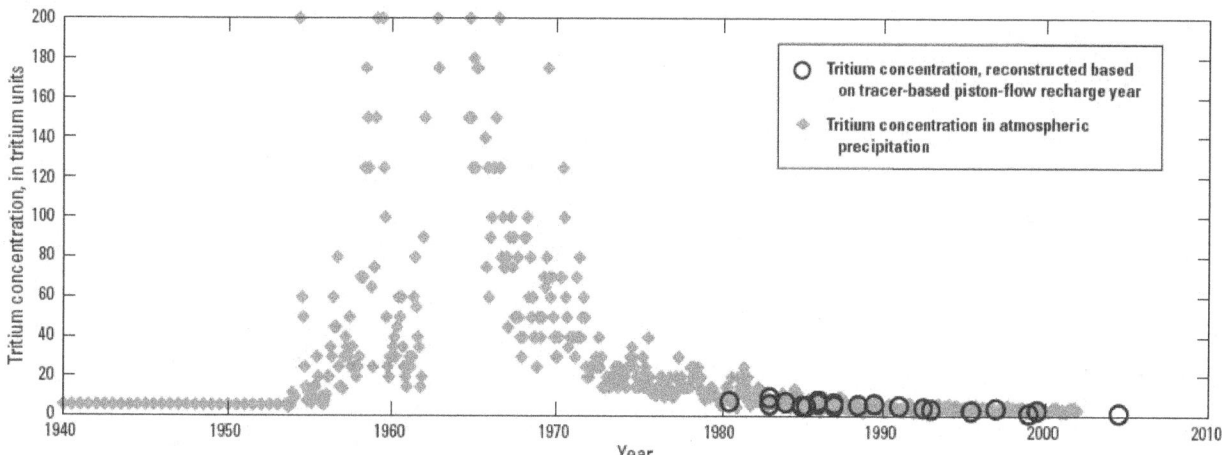

Figure B145.   Reconstructed tritium concentrations (using SF$_6$-based ages) and tritium in atmospheric precipitation, LUSRC1 network, SCTX Study Unit.

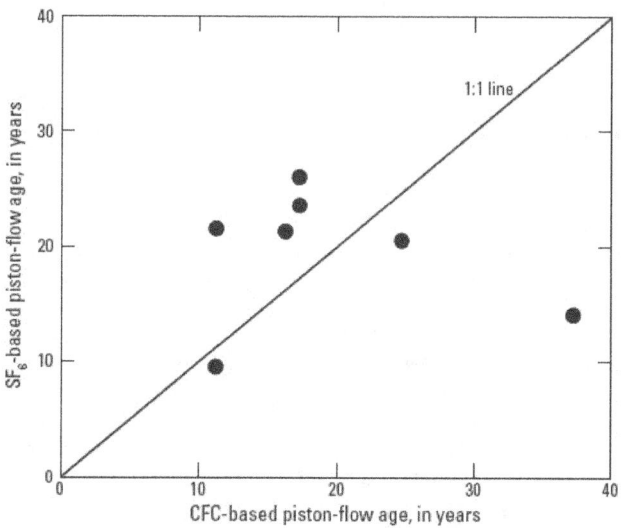

Figure B146.   SF$_6$- versus CFC-based age comparison, LUSRC1 network, SCTX Study Unit.

SCTX LUSRC2 and REFRE2

Samples from 24 sites in the SCTX Study Unit were collected in 2008 for CFCs, SF$_6$, and ^{3}H/^{3}He (networks and, in parentheses, number of sites):

. LUSRC2 (23)

. REFRE2 (1)

The aquifer is composed of limestone of the Edwards-Trinity aquifer system.

Major dissolved-gas data were available for 24 sites. Of these 24 sites, 20 were oxic and 4 were suboxic.

Age interpretations from tracer concentrations were made assuming that recharge elevation was equal to the elevation of the water table. Estimates of recharge temperature and excess air were based on major dissolved-gas data. For the suboxic sites, if the oxic sites were used to constrain the excess sir, the recharge temperatures were too low, so used the median recharge temperature of the oxic sites for the suboxic sites.

^{3}H/^{3}He ages were calculated for twelve sites (9 of the 12 sites required a correction for terrigenic helium), while 1 site was not datable because the tritium was too low, 7 sites were not datable because of fractionation, and samples from 4 sites were lost due to high pressure or damaged copper tubes.

The raw tracer data, major dissolved-gas data, the ancillary chemical and well construction data that were used in the interpretations, and the piston-flow ages are presented in table B42.

. Advantages associated with these samples:

. Multiple tracers (CFCs, SF$_6$, and ^{3}H/^{3}He, as well as major dissolved gases).

. Mixture of monitoring wells and one irrigation well, so likely low pumping stress.

. Disadvantages associated with these samples:

. Relatively large open intervals ranging from 23.77-100 feet so mixing likely.

. Median penetration of center of open interval into water table was 33.95 feet (not sampling close to the water table, potentially mixing).

. Depth to water (can affect tracer transport to water table):

. Median: 67.89 feet

. Mean: 76.60 feet

. Min: 16.58 feet

. Max: 167.99 feet

. Brief analysis:

. The CFC-, SF$_6$-, and ^{3}H/^{3}He-based age gradients for these sites are shown in figures B147, B148, and B149.

The age gradients show a great deal of scatter as would be expected for samples taken from wells finished in karst with large open intervals. The ^{3}H/^{3}He-based ages appear to be biased toward young ages, which may be the result of helium loss in karst environments.

The reconstructed ^{3}H plots for CFC-, SF$_6$-, and ^{3}H/^{3}He-based ages are shown in figures B150, B151, and B152. The reconstructions show evidence of mixing with old groundwater.

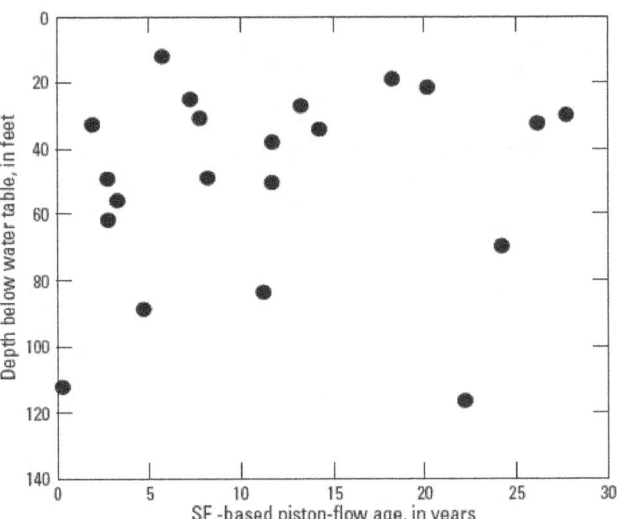

Figure B148. SF$_6$-based age gradient for dated sites from the LUSRC2 and REFRE2 networks, SCTX Study Unit.

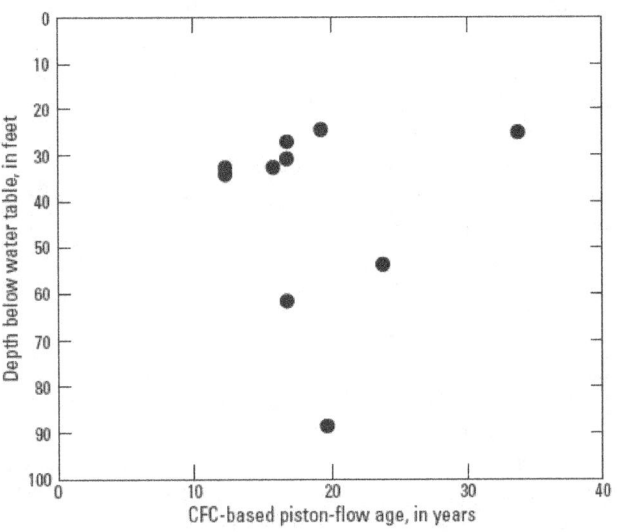

Figure B147. CFC-based age gradient for dated sites from the LUSRC2 and REFRE2 networks, SCTX Study Unit.

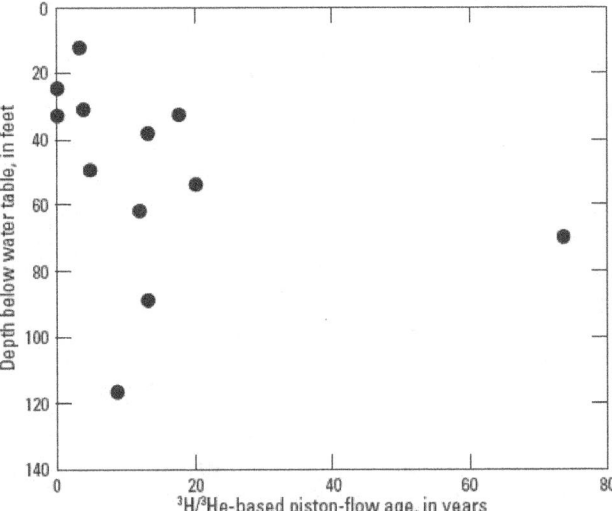

Figure B149. ^{3}H/^{3}He-based age gradient for dated sites from the LUSRC2 and REFRE2 networks, SCTX Study Unit.

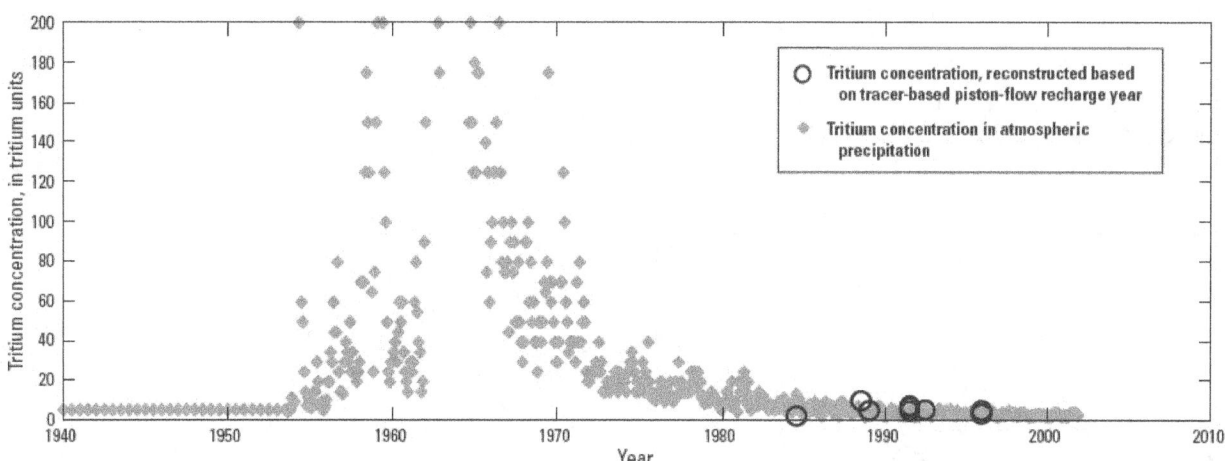

Figure B150. Reconstructed tritium concentrations (using CFC-based ages) and tritium in atmospheric precipitation, LUSRC2 and REFRE2 networks, SCTX Study Unit.

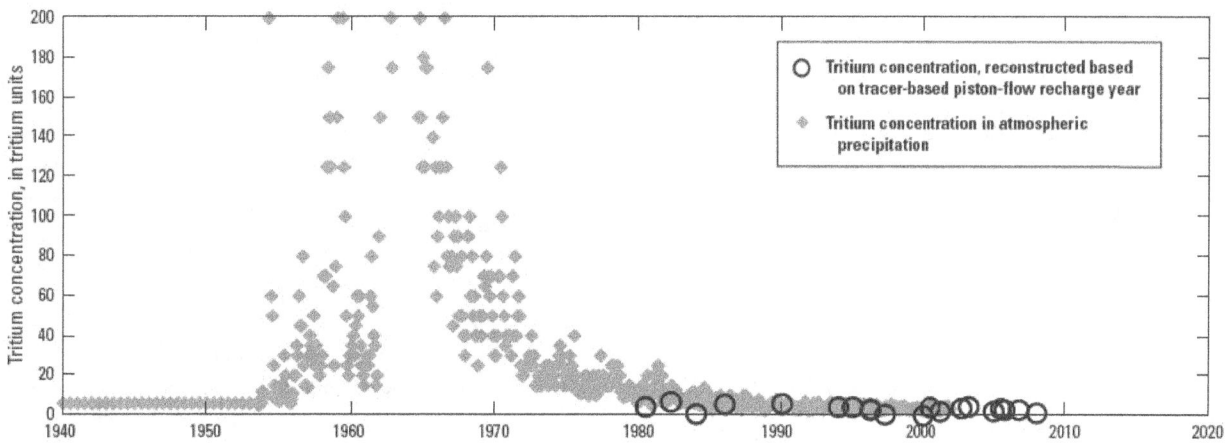

Figure B151. Reconstructed tritium concentrations (using SF$_6$-based ages) and tritium in atmospheric precipitation, LUSRC2 and REFRE2 networks, SCTX Study Unit.

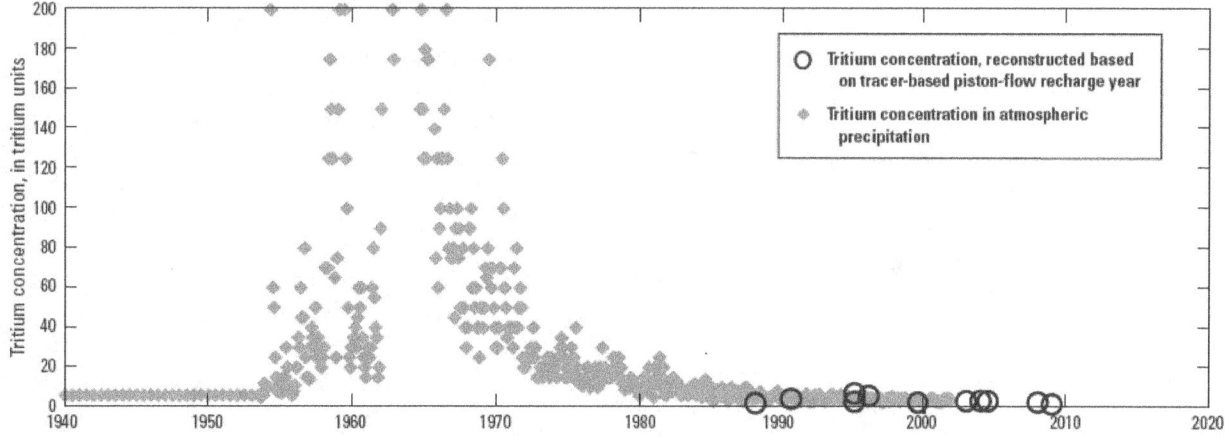

Figure B152. Reconstructed tritium concentrations (using ^{3}H/^{3}He-based ages) and tritium in atmospheric precipitation, LUSRC2 and REFRE2 networks, SCTX Study Unit.

The SF_6- versus CFC-based age comparison, the $^3H/^3He$- versus CFC-based age comparison, and the $^3H/^3He$- versus SF_6-based age comparison for this network are shown in figures B153, B154, and B155. The age comparisons show similar trends to the figures shown above, with a great deal of scatter resulting from mixing in karst environments.

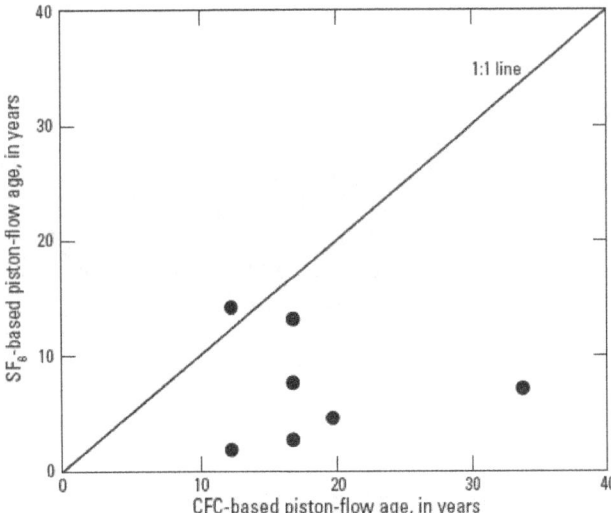

Figure B153. SF_6- versus CFC-based age comparison, LUSRC2 and REFRE2 networks, SCTX Study Unit.

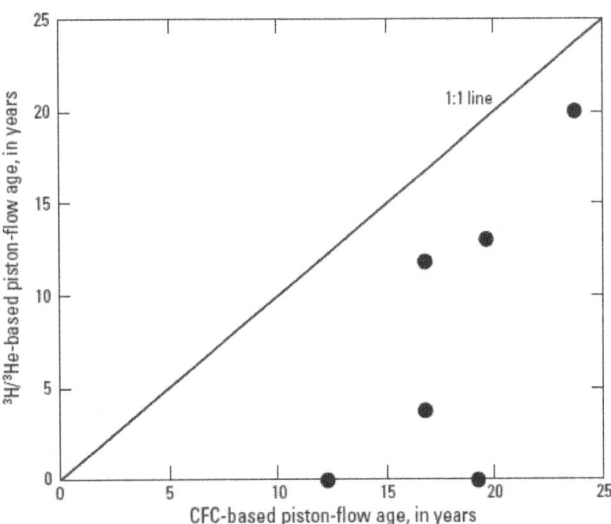

Figure B154. $^3H/^3He$- versus CFC-based age comparison, LUSRC2 and REFRE2 networks, SCTX Study Unit.

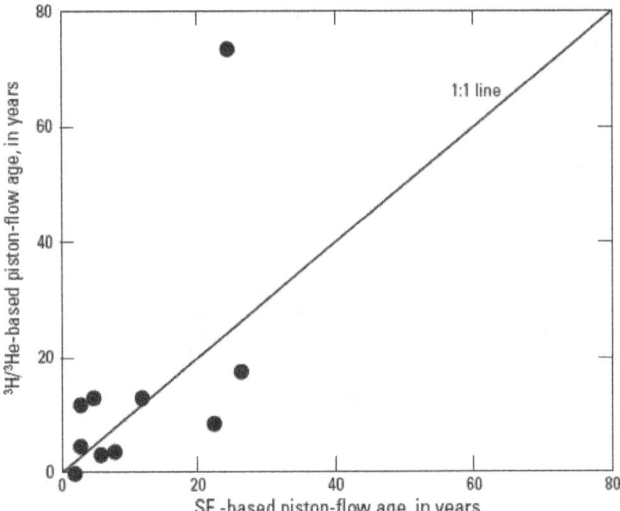

Figure B155. $^3H/^3He$- versus SF_6-based age comparison, LUSRC2 and REFRE2 networks, SCTX Study Unit.

SCTX SUS4

Samples from 30 sites in the SCTX Study Unit were collected in 2008 for CFCs, SF_6, and $^3H/^3He$ (networks and, in parentheses, number of sites):

.   SUS4 (CFCs, 29; SF_6, 28; $^3H/^3He$, 17)

The aquifer is composed of sands of the Texas coastal uplands aquifer system.

Major dissolved-gas data were available for 30 sites. Of these 30 sites, 4 were oxic and 26 were suboxic. Using the oxic sites to constrain the excess air for the suboxic sites resulted in unreasonable recharge temperatures. The median recharge temperature of the dissolved-gas data was 19.3, which was similar to the MAAT+1°C, so used MAAT+1°C.

Age interpretations from tracer concentrations were made assuming that recharge elevation was equal to the elevation of the water table, that recharge temperature was equal to the mean annual air temperature +1°C, and that excess air concentrations were 2 cc STP/kg.

$^3H/^3He$ ages were calculated for eight sites (only one of the eight sites required a correction for terrigenic helium).

The raw tracer data, major dissolved-gas data, the ancillary chemical and well construction data that were used in the interpretations, and the piston-flow ages are presented in table B43.

.   Advantages associated with these samples:

.   Multiple tracers (CFCs, SF_6, $^3H/^3He$, as well as major dissolved gases).

Disadvantages associated with these samples:

- Mixture of domestic, irrigation, stock, and public supply wells, so variable pumping rates.

- Relatively large open intervals ranging from 4 to 650 feet so mixing likely.

- Median penetration of center of open interval into water table was 212.9 feet (not sampling close to the water table, potentially mixing).

- Suboxic conditions.

Depth to water (can affect tracer transport to water table):

- Median: 208.61 feet

- Mean: 218.36 feet

- Min: 7.36 feet

- Max: 619.06 feet

Brief analysis:

The SF_6-based age gradient for these sites is shown in figure B156. The age gradient shows a great deal of scatter as would be expected for samples taken from such a wide variety of types of wells with large open intervals. Differences in screen length, recharge source/ strength, aquifer heterogeneity, pumping stresses, and the position of the well within the flow system may cause some wells to deviate from the general pattern of increasing age with depth. In addition, the suboxic conditions could be stripping the SF_6 from the water resulting in ages that are biased old.

The reconstructed ^3H plots for SF_6- and $^3H/^3He$-based ages are shown in figures B157 and B158. The reconstructions show evidence of mixing and/or SF_6 loss due to suboxic conditions.

The $^3H/^3He$- versus SF_6-based age comparison for this network is shown in figure B159. The age comparison is poor and likely results from mixing and possible gas stripping of SF_6 due to suboxic conditions.

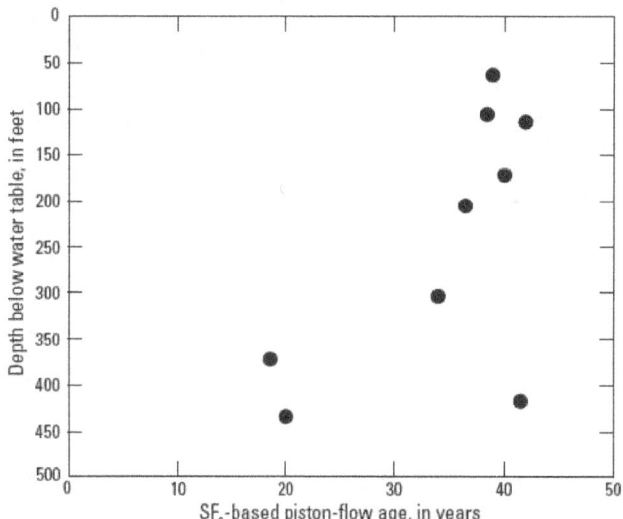

Figure B156. SF_6-based age gradient for dated sites from the SUS4 network, SCTX Study Unit.

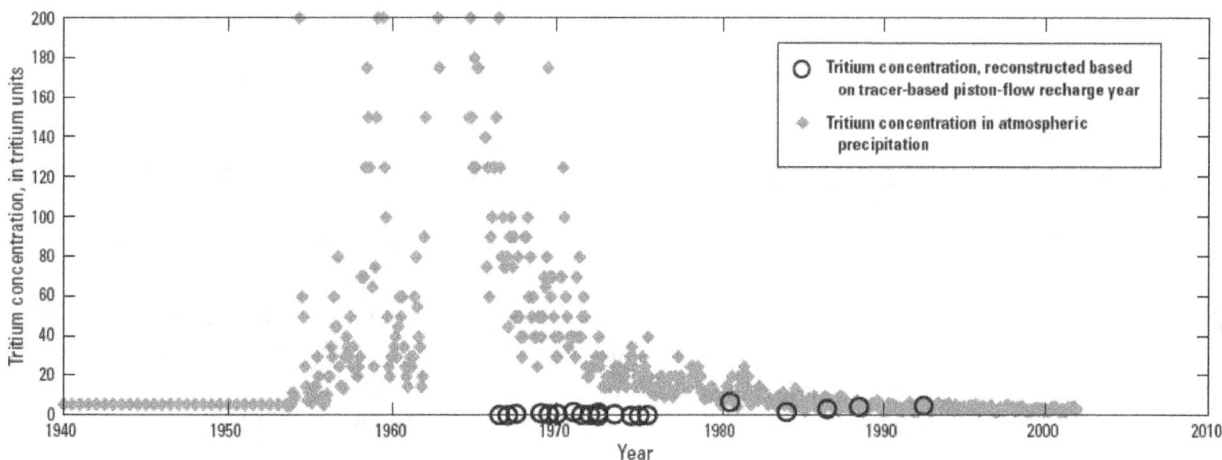

Figure B157. Reconstructed tritium concentrations (using SF_6-based ages) and tritium in atmospheric precipitation, SUS4 network, SCTX Study Unit.

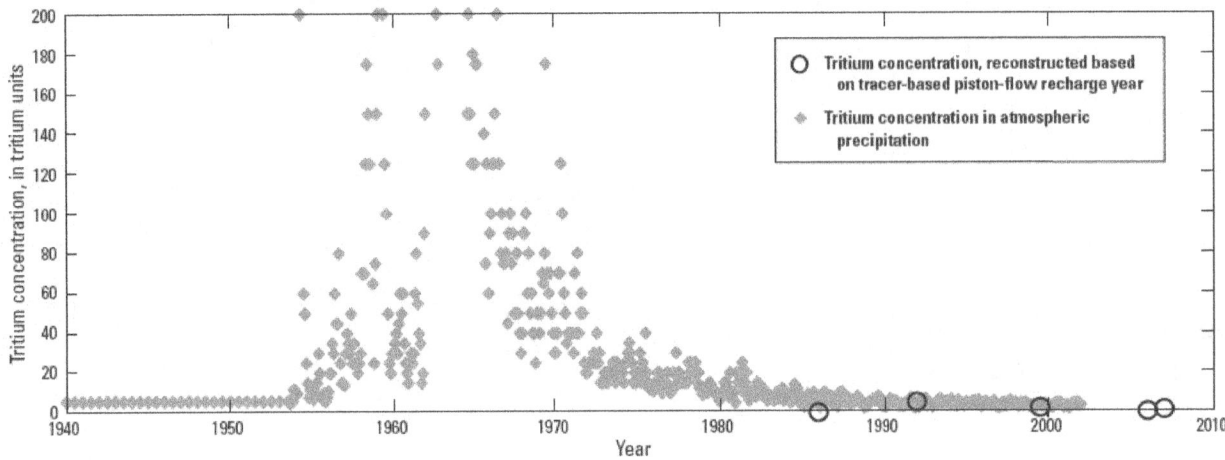

Figure B158. Reconstructed tritium concentrations (using ^{3}H/^{3}He-based ages) and tritium in atmospheric precipitation, SUS4 network, SCTX Study Unit.

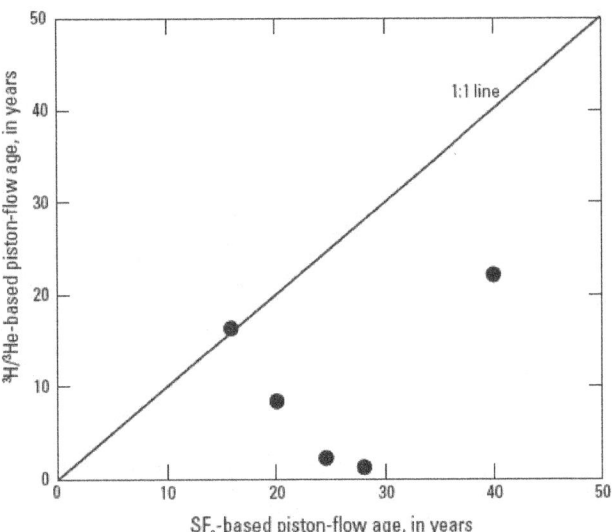

Figure B159. ^{3}H/^{3}He- versus SF$_6$-based age comparison, SUS4 network, SCTX Study Unit.

SOCA SUS1 and SUS2

Samples from 7 sites in the SOCA Study Unit were collected in 2008 for SF$_6$ and ^{3}H/^{3}He (networks and, in parentheses, number of sites):

. SUS1 (SF$_6$, 4)

. SUS2 (SF$_6$, 3; ^{3}H/^{3}He, 3)

The aquifer is composed of sand, gravel, silt, and clay. Major dissolved-gas data were available for 10 sites. Of these 10 sites, 8 were oxic and 2 were suboxic.

Age interpretations from tracer concentrations were made assuming that recharge elevation was equal to the elevation of the water table. Estimates of recharge temperature and excess air were based on major dissolved-gas data, with recharge temperature and excess air at suboxic sites being constrained using median excess air at oxic sites.

^{3}H/^{3}He ages were calculated for two sites (both sites required a correction for terrigenic helium), while one site was not datable because the tritium was too low.

The raw tracer data, major dissolved-gas data, the ancillary chemical and well construction data that were used in the interpretations, and the piston-flow ages are presented in table B44.

. Advantages associated with these samples:

. Multiple tracers (SF$_6$ and ^{3}H/^{3}He, as well as major dissolved gases).

. Disadvantages associated with these samples:

. Public supply wells, so variable pumping rates more likely.

. Relatively large open intervals ranging from 169 to 931 feet so mixing likely.

. Median penetration of center of open interval into water table was 371.2 feet (not sampling close to the water table, potentially mixing).

. Depth to water (can affect tracer transport to water table):

. Median: 130.05 feet

. Mean: 164.50 feet

. Min: 46.00 feet

. Max: 360.00 feet

. Brief analysis:

The SF$_6$-based age gradient for these sites is shown in figure B160. The age gradient shows a great deal of scatter as would be expected for samples taken from wells with large open intervals. Differences in screen length, recharge source/strength, aquifer heterogeneity, pumping stresses, and the position of the well within the flow system may cause some wells to deviate from the general pattern of increasing age with depth.

The reconstructed ^{3}H plot for SF$_6$-based ages is shown in figure B161. The reconstruction shows evidence of mixing as would be expected for such a wide variety of pumping rates and large open intervals.

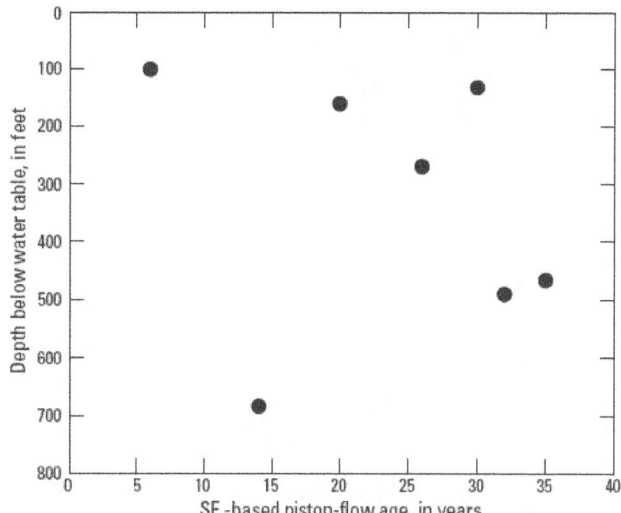

Figure B160. SF$_6$-based age gradient for dated sites from the SUS1 and SUS2 networks, SOCA Study Unit.

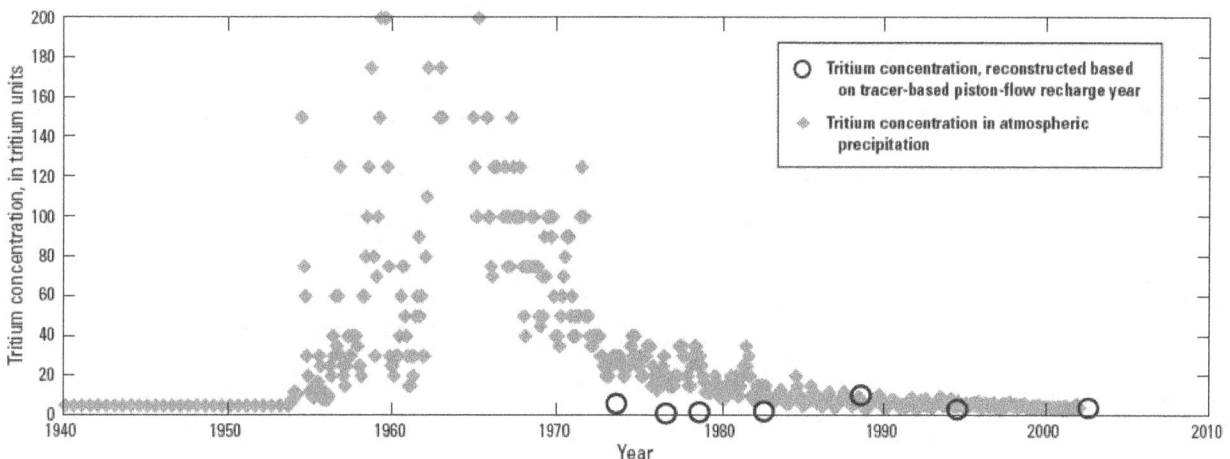

Figure B161. Reconstructed tritium concentrations (using SF$_6$-based ages) and tritium in atmospheric precipitation, SUS1 and SUS2 networks, SOCA Study Unit.

SOFL LUSOR1 and LUSOT1

Samples from 13 sites in the SOFL Study Unit were collected in 2009 for CFCs, SF$_6$, and ^{3}H/^{3}He (networks and, in parentheses, number of sites):

. LUSOR1 (CFCs, 11; SF$_6$, 8; ^{3}H/^{3}He, 11)

. LUSOT1 (SF$_6$, 2)

The aquifer is composed of sand.

Major dissolved-gas data were available for nine LUSOR1 sites. Of these nine sites, all were suboxic.

Age interpretations from tracer concentrations were made assuming that recharge elevation was equal to the elevation of the water table, that recharge temperature was equal to the mean annual air temperature +1°C, and that excess air concentrations were 2 cc STP/kg.

^{3}H/^{3}He ages could not be calculated for any sites as a result of fractionation.

The raw tracer data, major dissolved-gas data, the ancillary chemical and well construction data that were used in the interpretations, and the piston-flow ages are presented in table B45.

. Advantages associated with these samples:

. Multiple tracers (CFC, SF$_6$, and ^{3}H/^{3}He, as well as major dissolved gases).

. Monitoring wells, therefore low pumping stress.

. Relatively short open intervals ranging from 2.95 to 10 feet so mixing likely minimized.

.   Median penetration of center of open interval into water table was 5.34 feet (sampling close to the water table, potentially minimizes mixing).

.   Disadvantages associated with these samples:

.   Suboxic conditions.

.   Depth to water (can affect tracer transport to water table):

.   Median: 3.69 feet

.   Mean: 4.59 feet

.   Min: 1.25 feet

.   Max: 9.05 feet

.   Brief analysis:

.   The SF_6-based age gradient for these sites is shown in figure B162. The age gradient shows a great deal of scatter as would be expected for samples taken from wells in suboxic conditions. The SF_6 concentrations have likely been affected by stripping due to high methane concentrations. In addition, differences in screen length, recharge source/strength, aquifer heterogeneity, pumping stresses, and the position of the well within the flow system may cause some wells to deviate from the general pattern of increasing age with depth.

The reconstructed ^3H plot for SF_6-based ages is shown in figure B163. The reconstruction shows evidence of unmixed, piston-flow transport, however, it is also likely that the SF_6 ages should be younger and would shift to the right on the ^3H input function.

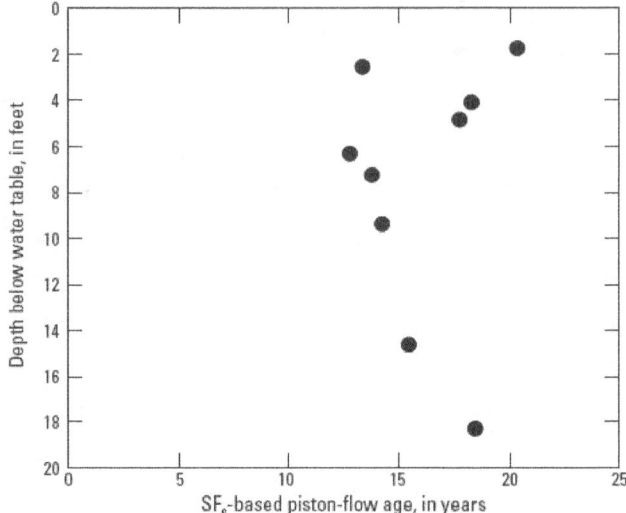

Figure B162. SF_6-based age gradient for dated sites from the LUSOR1 and LUSOT1 networks, SOFL Study Unit.

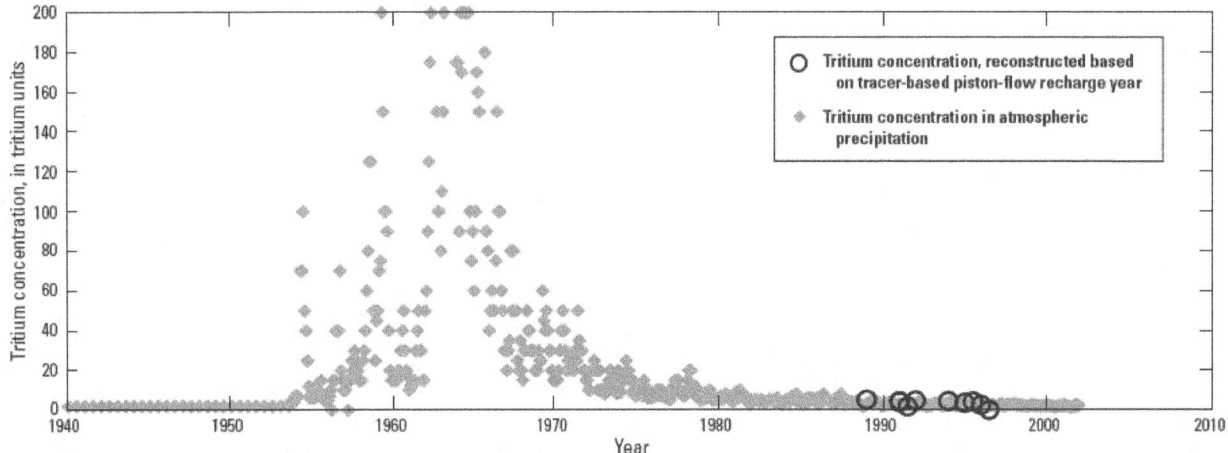

Figure B163.   Reconstructed tritium concentrations (using SF_6-based ages) and tritium in atmospheric precipitation, LUSOR1 and LUSOT1 networks, SOFL Study Unit.

SOFL LUSRC1a

Samples from 4 sites in the SOFL Study Unit were collected in 2009 for SF_6 and $^3H/^3He$, and 21 sites in the SOFL Study Unit were collected in 2010 for SF_6 (networks and, in parentheses, number of sites):

.   LUSRC1a (SF_6, 4 in 2009 and 21 in 2010; $^3H/^3He$, 3 in 2009)

The aquifer is composed of limestone of the Biscayne Limestone aquifer.

Major dissolved-gas data were available for 21 sites. Of these 21 sites, all were suboxic, and some had very high methane concentrations.

Age interpretations from tracer concentrations were made assuming that recharge elevation was equal to the elevation of the water table, that recharge temperature was equal to the mean annual air temperature +1°C, ad that excess air concentrations were 2 cc STP/kg.

$^3H/^3He$ ages could not be calculated for any sites as a result of fractionation.

The raw tracer data, major dissolved-gas data, the ancillary chemical and well construction data that were used in the interpretations, and the piston-flow ages are presented in table B46.

. Advantages associated with these samples:

. Multiple tracers (SF_6 and $^3H/^3He$, as well as major dissolved gases).

. Monitoring wells, therefore low pumping stress.

. Relatively short open intervals ranging from 2-10 feet so mixing likely minimized.

. Median penetration of center of open interval into water table was 9.25 feet (sampling close to the water table, potentially minimizes mixing).

. Disadvantages associated with these samples:

. Suboxic conditions.

. Depth to water (can affect tracer transport to water table):

. Median: 3.82 feet

. Mean: 4.46 feet

. Min: 1.86 feet

. Max: 13.60 feet

. Brief analysis:

. The SF_6-based age gradient for these sites is shown in figure B164. The age gradient shows a great deal of scatter as would be expected for samples taken from wells in suboxic conditions. The SF_6 concentrations have likely been affected by stripping due to high

methane concentrations. In addition, differences in screen length, recharge source/strength, aquifer heterogeneity, pumping stresses, and the position of the well within the flow system may cause some wells to deviate from the general pattern of increasing age with depth.

The reconstructed ^3H plot for SF_6-based ages is shown in figure B165. The reconstruction shows evidence of unmixed, piston-flow transport, however, it is also likely that the SF_6 ages should be younger and would shift to the right on the ^3H input function.

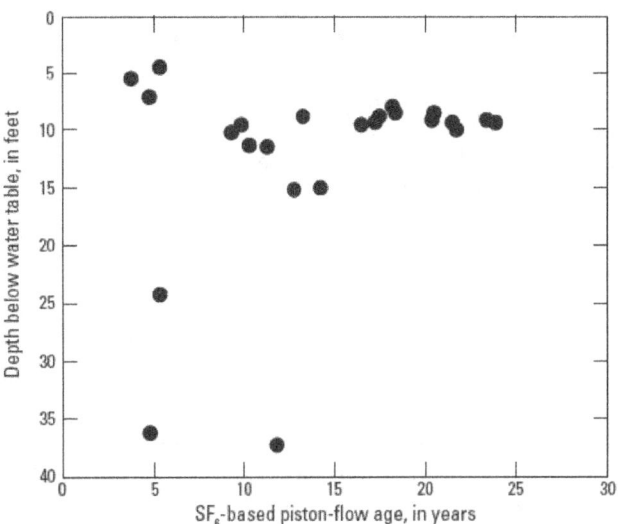

Figure B164. SF_6-based age gradient for dated sites from the LUSRC1a network, SOFL Study Unit.

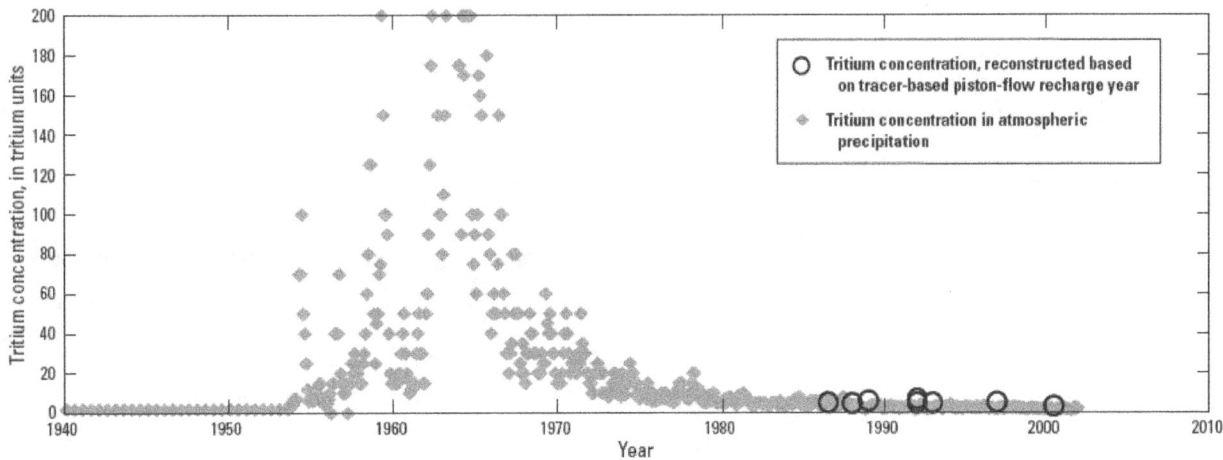

Figure B165. Reconstructed tritium concentrations (using SF_6-based ages) and tritium in atmospheric precipitation, LUSRC1a network, SOFL Study Unit.

SOFL SUS1

Samples from five sites in the SOFL Study Unit were collected in 2009 for SF_6 and $^3H/3He$ (networks and, in parentheses, number of sites):

.   SUS1 (5)

The aquifer is composed of limestone of the Biscayne Limestone aquifer.

Major dissolved-gas data were available for five sites. Of these five sites, all five were suboxic.

Age interpretations from tracer concentrations were made assuming that recharge elevation was equal to the elevation of the water table, that recharge temperature was equal to the mean annual air temperature $+1°C$, and that excess air concentrations were 2 cc STP/kg.

$^3H/^3He$ ages could not be calculated for any sites as a result of fractionation.

The raw tracer data, major dissolved-gas data, the ancillary chemical and well construction data that were used in the interpretations, and the piston-flow ages are presented in table B47.

.   Advantages associated with these samples:

        Multiple tracers (SF_6 and $^3H/^3He$, as well as major dissolved gases).

.   Disadvantages associated with these samples:

        Mixture of monitoring and public supply wells, so variable pumping rates.

.   Relatively large open intervals ranging from 7 to 50 feet (with 2 unknown) so mixing likely.

.   Median penetration of center of open interval into water table was 68.37 feet (not sampling close to the water table, potentially mixing).

.   Suboxic conditions.

.   Fractionation of $^3H/^3He$ samples.

.   Depth to water (can affect tracer transport to water table):

.   Median: 5.30 feet

.   Mean: 9.64 feet

.   Min: 3.00 feet

.   Max: 19.13 feet

.   Brief analysis:

        The reconstructed ^3H plot for SF_6-based ages is shown in figure B166. The reconstruction shows evidence of unmixed, piston-flow transport, however, it is also likely that the SF_6 ages should be younger and would shift to the right on the ^3H input function.

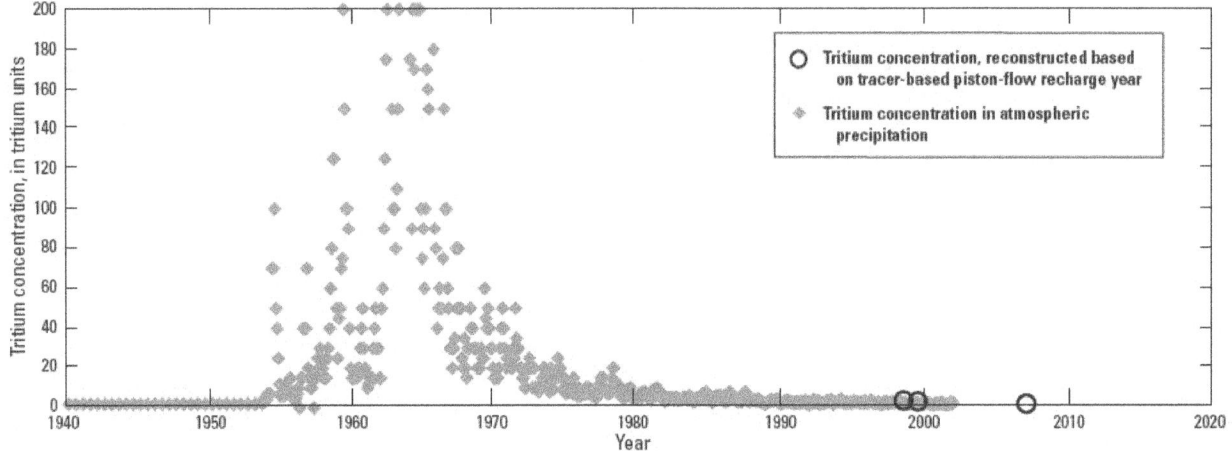

Figure B166. Reconstructed tritium concentrations (using SF_6-based ages) and tritium in atmospheric precipitation, SUS1 network, SOFL Study Unit.

SPLT LUSCR1

Samples from four sites in the SPLT Study Unit were collected in 2008 for CFCs and $^3H/^3He$ (networks and, in parentheses, number of sites):

. LUSCR1 (4)

The aquifer is composed of alluvial gravel, sand, and silt. Major dissolved-gas data were available for all four sites. Of these four sites, all four were oxic, however, because recharge temperatures and excess air were so variable, they were not used in the CFC and $^3H/^3He$ spreadsheets.

Age interpretations from tracer concentrations were made assuming that recharge elevation was equal to the elevation of the water table, that recharge temperature was equal to the mean annual air temperature +1°C, and that excess air concentrations were 2 cc STP/kg.

$^3H/^3He$ ages were calculated for all four sites and did not require a correction for terrigenic He.

The raw tracer data, major dissolved-gas data, the ancillary chemical and well construction data that were used in the interpretations, and the piston-flow ages are presented in table B48.

. Advantages associated with these samples:

.   Multiple tracers (CFCs and $^3H/^3He$, as well as major dissolved gases).

.   Monitoring wells, therefore low pumping stress.

.   Relatively short open intervals ranging from 7 to 10.78 feet so mixing likely minimized.

.   Median penetration of center of open interval into water table was 5.05 feet (sampling close to the water table, potentially minimizing mixing).

. Disadvantages associated with these samples:

.   None.

. Depth to water (can affect tracer transport to water table):

.   Median: 7.21 feet

.   Mean: 13.02 feet

.   Min: 3.79 feet

.   Max: 33.86 feet

. Brief analysis:

.   The CFC- and $^3H/^3He$-based age gradients for these sites are shown in figures B167 and B168. The age gradients show a great deal of scatter as would be expected for samples taken from such a narrow range in depth and in a cropland area affected by irrigation. The $^3H/^3He$-based ages are also significantly younger than the CFC-based ages indicating likely helium loss during irrigation. In addition, differences in screen length, recharge source/strength, aquifer heterogeneity, pumping stresses, and the position of the well within the flow system may cause some wells to deviate from the general pattern of increasing age with depth.

The reconstructed ^3H plots for CFC- and $^3H/^3He$-based ages are shown in figures B169 and B170. The reconstructions show evidence of unmixed, piston-flow transport, however, as with the age gradient plots above, there is a significant discrepancy between CFC- and $^3H/^3He$-based ages, likely resulting from irrigation practices.

The $^3H/^3He$- versus CFC-based age comparison for this network is shown in figure B171. The age comparison is limited by the low number of samples, but also shows poor agreement likely resulting from helium loss during irrigation.

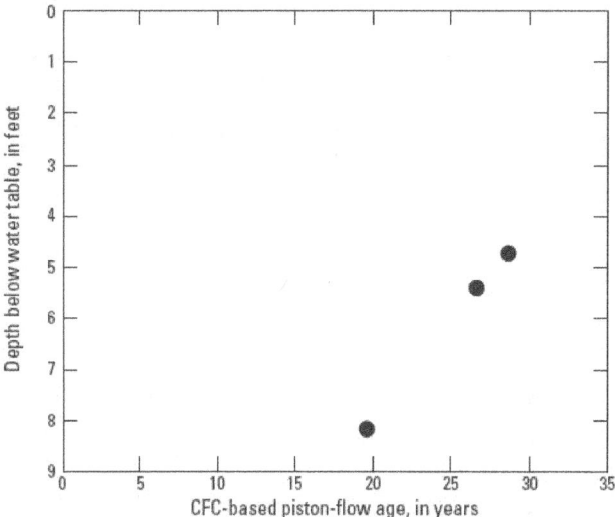

Figure B167.   CFC-based age gradient for dated sites from the LUSCR1 network, SPLT Study Unit.

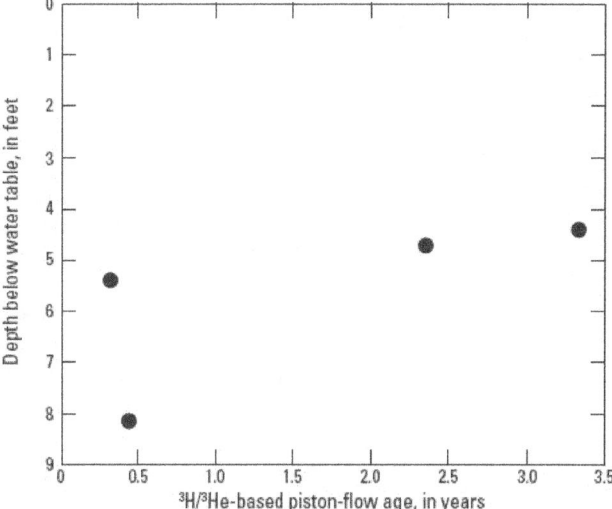

Figure B168. $^3H/^3He$-based age gradient for dated sites from the LUSCR1 network, SPLT Study Unit.

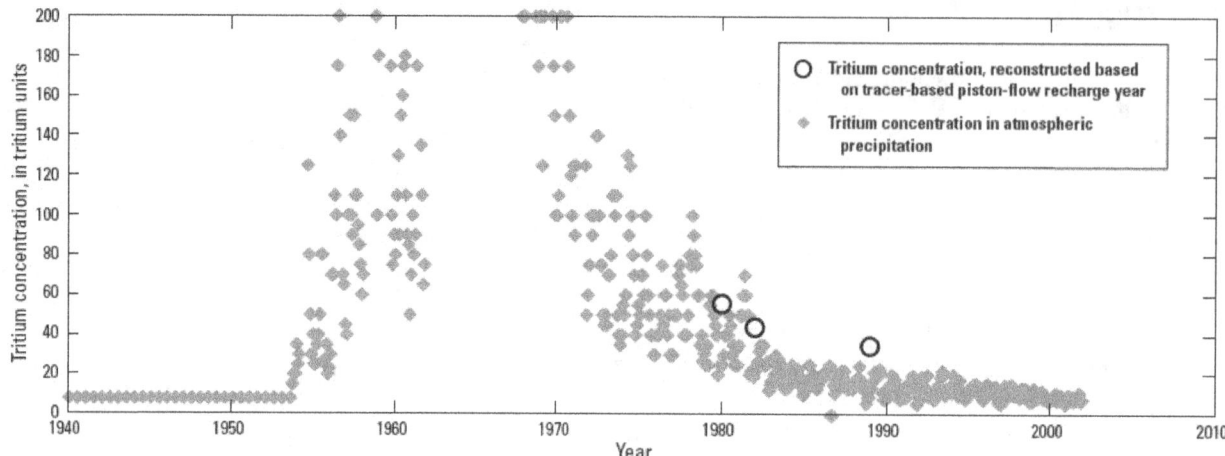

Figure B169.   Reconstructed tritium concentrations (using CFC-based ages) and tritium in atmospheric precipitation, LUSCR1 network, SPLT Study Unit.

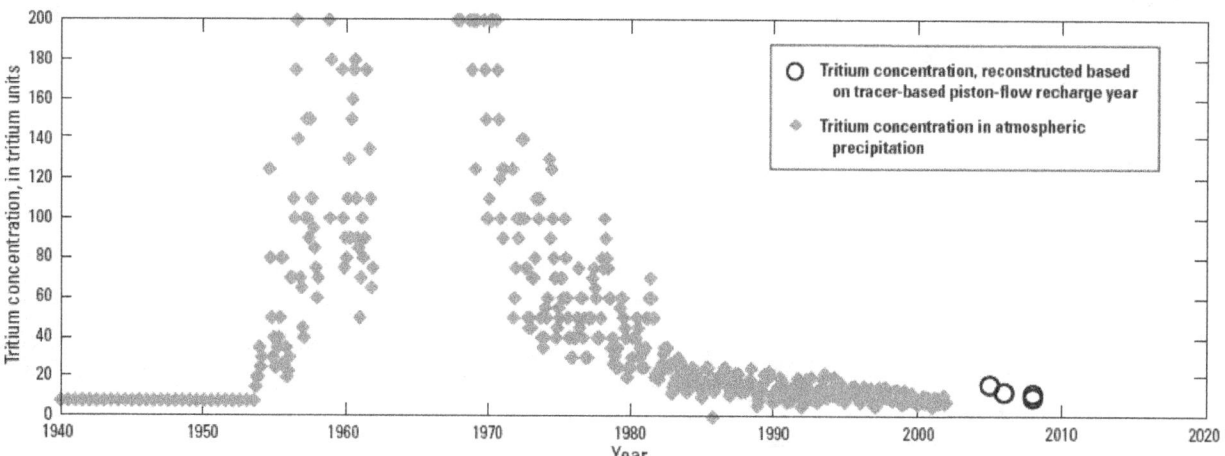

Figure B170.   Reconstructed tritium concentrations (using ^{3}H/^{3}He-based ages) and tritium in atmospheric precipitation, LUSCR1 network, SPLT Study Unit.

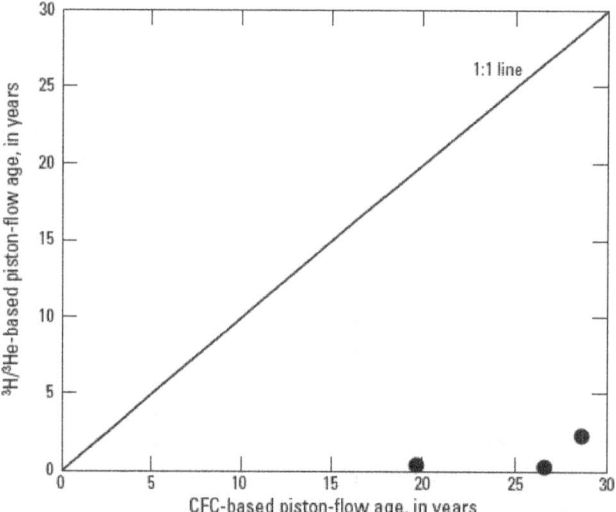

Figure B171. ^{3}H/^{3}He- versus CFC-based age comparison, LUSCR1 network, SPLT Study Unit.

SPLT LUSRC2

Samples from five sites (two of which were sampled twice) in the SPLT Study Unit were collected in 2007 for CFCs and $^3H/^3He$ (networks and, in parentheses, number of sites):

.   LUSRC2 (5)

The aquifer is composed of sandstone of the Dawson Arkose for three of the sites, and alluvium for two of the sites.

Major dissolved-gas data were available for all five sites. Of these five sites, two were oxic and three were suboxic. Recharge temperatures ranged from 5 to 18°C and were even lower if denitrification was assumed.

Age interpretations from tracer concentrations were made assuming that recharge elevation was equal to the elevation of the water table, that recharge temperature was equal to the mean annual air temperature +1°C, and that excess air concentrations were 2 cc STP/kg.

$^3H/^3He$ ages were calculated for four sites (only one of the four sites required a correction for terrigenic helium), while one site was not datable because of fractionation.

The raw tracer data, major dissolved-gas data, the ancillary chemical and well construction data that were used in the interpretations, and the piston-flow ages are presented in table B49.

.   Advantages associated with these samples:

.   Multiple tracers (CFCs and $^3H/^3He$, as well as major dissolved gases).

.   Monitoring wells, therefore low pumping stress.

.   Relatively short open intervals <10 feet so mixing likely minimized.

.   Median penetration of center of open interval into water table was 11.66 feet (sampling close to the water table, potentially minimizes mixing).

.   Disadvantages associated with these samples:

.   None.

.   Depth to water (can affect tracer transport to water table):

.   Median: 33.83 feet

.   Mean: 28.43 feet

.   Min: 10.95 feet

.   Max: 41.46 feet

.   Brief analysis:

.   The CFC- and $^3H/^3He$-based age gradients for these sites are shown in figures B172 and B173. The age gradients show little structure, likely as a result of the narrow range in depths for these wells. The $^3H/^3He$-based ages also are significantly younger than the CFC-based ages indicating likely helium loss.

The reconstructed ^3H plots for CFC- and $^3H/^3He$-based ages are shown in figures B174 and B175. The reconstructions show evidence of unmixed, piston-flow transport, however, as with the age gradient plots above, there is a significant discrepancy between CFC- and $^3H/^3He$-based ages, likely resulting from irrigation practices near the residential and commercial areas.

The $^3H/^3He$- versus CFC-based age comparison for this network is shown in figure B176. The age comparison is poor, likely as a result of irrigation practices near the residential and commercial areas.

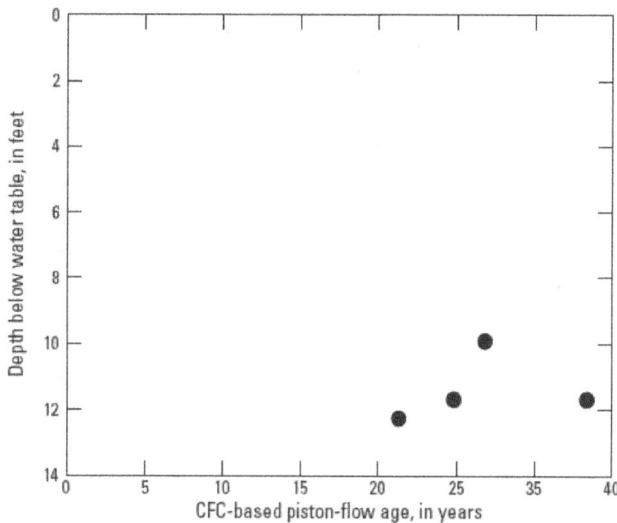

Figure B172.   CFC-based age gradient for dated sites from the LUSCR2 network, SPLT Study Unit.

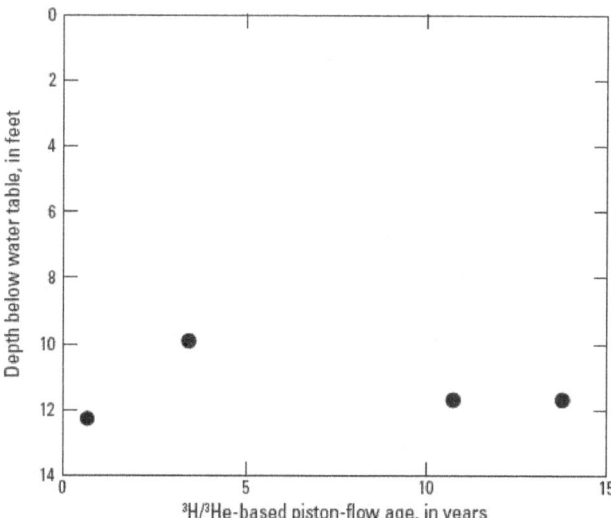

Figure B173. $^3H/^3He$-based age gradient for dated sites from the LUSCR2 network, SPLT Study Unit.

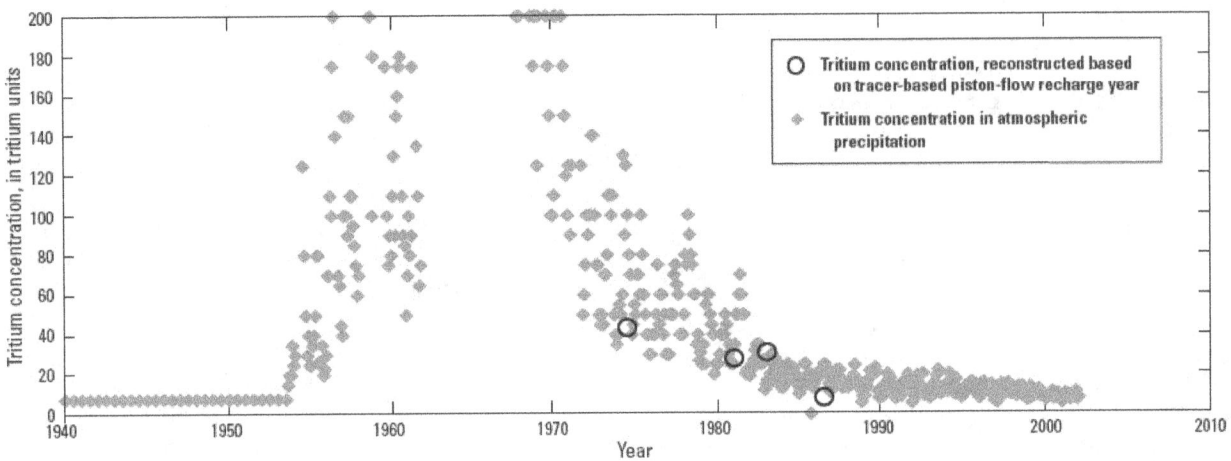

Figure B174. Reconstructed tritium concentrations (using CFC-based ages) and tritium in atmospheric precipitation, LUSRC2 network, SPLT Study Unit.

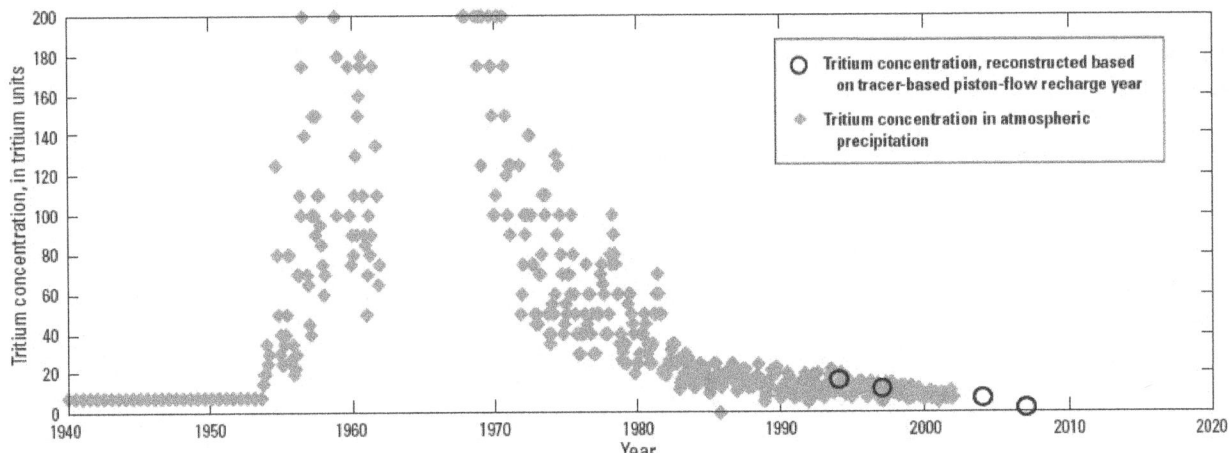

Figure B175. Reconstructed tritium concentrations (using ^{3}H/^{3}He-based ages) and tritium in atmospheric precipitation, LUSRC2 network, SPLT Study Unit.

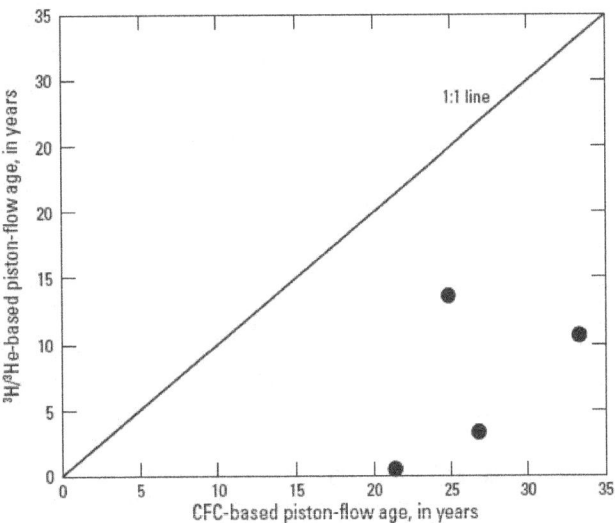

Figure B176. ^{3}H/^{3}He- versus CFC-based age comparison, LUSRC2 network, SPLT Study Unit.

TRIN LUSRC1 and REFOT1

Samples from six sites in the TRIN Study Unit were collected in 2007 for CFCs and SF$_6$ (networks and, in parentheses, number of sites):

.   LUSRC1 (3)

.   REFOT1 (3)

The aquifer is composed of sand and clay of the Chicot Aquifer.

Major dissolved-gas data were available for all six sites. Of these six sites, four were oxic and two were suboxic.

Age interpretations from tracer concentrations were made assuming that recharge elevation was equal to the elevation of the water table. Estimates of recharge temperature and excess air were based on major dissolved-gas data, with recharge temperature and excess air at suboxic sites being constrained using median excess air at oxic sites.

The raw tracer data, major dissolved-gas data, the ancillary chemical and well construction data that were used in the interpretations, and the piston-flow ages are presented in table B50.

- Advantages associated with these samples:

  - Multiple tracers (CFCs and SF$_6$, as well as major dissolved gases).

  - Monitoring wells, therefore low pumping stress.

  - Relatively short open intervals of 10 feet so mixing likely minimized.

  - Median penetration of center of open interval into water table was 10.7 feet (sampling close to the water table, potentially minimizes mixing).

- Disadvantages associated with these samples:

  - Suboxic conditions.

  - No tritium.

- Depth to water (can affect tracer transport to water table):

  - Median: 14.31 feet

  - Mean: 12.63 feet

  - Min: 0.89 feet

  - Max: 25.91 feet

- Brief analysis:

  - The CFC- and SF$_6$-based age gradients for these sites are shown in figures B177 and B178. The age gradients show a great deal of scatter and the CFC-based ages

are older than the SF$_6$-based ages. The older CFC-based ages are likely the result of suboxic conditions in the aquifer. Differences in screen length, recharge source/strength, aquifer heterogeneity, pumping stresses, and the position of the well within the flow system may cause some wells to deviate from the general pattern of increasing age with depth.

The SF$_6$- versus CFC-based age comparison for this network is shown in figure B179. The age comparison is poor, likely as the result of CFC degradation in the suboxic conditions in the aquifer.

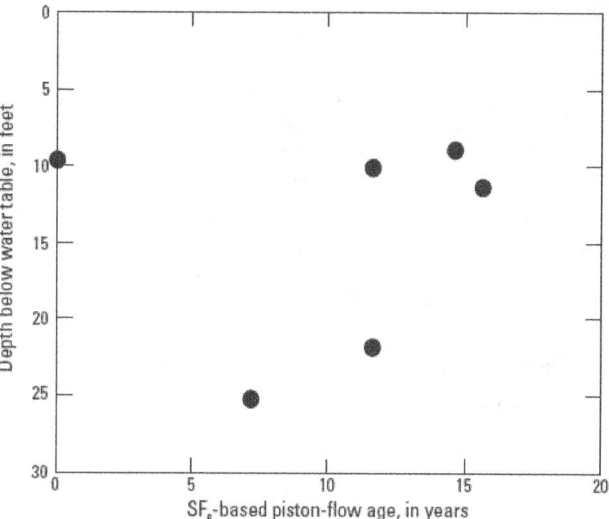

Figure B178. SF$_6$-based age gradient for dated sites from the LUSRC1 and REFOT1 networks, TRIN Study Unit.

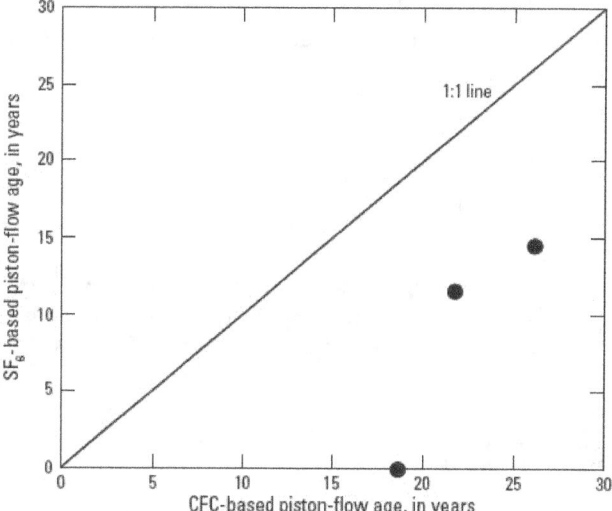

Figure B179. SF$_6$- versus CFC-based age comparison, LUSRC1 and REFOT1 networks, TRIN Study Unit.

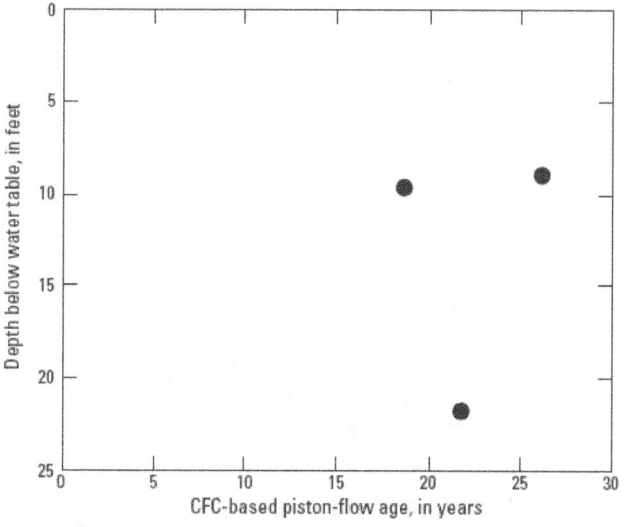

Figure B177. CFC-based age gradient for dated sites from the LUSRC1 and REFOT1 networks, TRIN Study Unit.

TRIN SUS3

Samples from five sites in the TRIN Study Unit were collected in 2007 for $^3H/^3He$ (networks and, in parentheses, number of sites):

.   SUS3 (5)

The aquifer is composed of sand and clay of the Chicot and Evangeline aquifers.

Major dissolved-gas data were not available.

Age interpretations from tracer concentrations were made assuming that recharge elevation was equal to the elevation of the water table, that recharge temperature was equal to the mean annual air temperature $+1°C$, and that excess air concentrations were 2 cc STP/kg.

$^3H/^3He$ ages could not be calculated for any sites. Four of the sites had tritium concentrations that were too low, and the sample from one site was lost due to high pressure (in addition to having a tritium concentration that was too low).

The raw tracer data, the ancillary chemical and well construction data that were used in the interpretations, and the piston-flow ages are presented in table B51.

.   Advantages associated with these samples:

.   None.

.   Disadvantages associated with these samples:

.   No multiple tracers or major dissolved gases.

.   Mixture of domestic and public supply wells, so variable pumping rates.

.   Relatively large open intervals ranging from 10 to 120 feet so mixing likely.

.   Median penetration of center of open interval into water table was 130.62 feet (not sampling close to the water table, potentially mixing).

.   Depth to water (can affect tracer transport to water table):

.   Median: 102.90 feet

.   Mean: 93.24 feet

.   Min: 30.38 feet

.   Max: 161.26 feet

.   Brief analysis:

.   The only thing that can be concluded from the $^3H/^3He$ data is that with tritium concentrations so low, the age of the water is pre-bomb, or greater than 54 years old.

UMIS FPSUR1

Samples from 16 sites in the UMIS Study Unit were collected in 2008 for CFCs (networks and, in parentheses, number of sites):

.   FPSUR1 (16)

The aquifer is composed of glacial sand, silt, clay, and gravel.

Major dissolved-gas data were available for all 16 sites.

Age interpretations from tracer concentrations were made assuming that recharge elevation was equal to the elevation of the water table, that recharge temperature was equal to the mean annual air temperature $+1°C$, and that excess air concentrations were 2 cc STP/kg.

The raw tracer data, major dissolved-gas data, the ancillary chemical and well construction data that were used in the interpretations, and the piston-flow ages are presented in table B52.

.   Advantages associated with these samples:

.   CFCs, as well as major dissolved gases.

.   Monitoring wells, therefore low pumping stress.

.   Relatively short open intervals ranging from 0.5 to 5 feet so mixing likely minimized.

.   Median penetration of center of open interval into water table was 7.10 feet (sampling close to the water table, potentially minimizes mixing).

.   Disadvantages associated with these samples:

.   Suboxic conditions.

.   Depth to water (can affect tracer transport to water table):

.   Median: 9.38 feet

.   Mean: 10.56 feet

.   Min: 1.00 feet

.   Max: 29.86 feet

.   Brief analysis:

.   Age-dating in this network was not possible as a result of the suboxic conditions and the fact that the only tracer samples that were taken were for CFCs, which were affected by degradation.

UMIS LUSCR1

Samples from 26 sites in the UMIS Study Unit were collected in 2006 for SF_6 and $^3H/^3He$ (networks and, in parentheses, number of sites):

.   LUSCR1 (SF_6, 26; $^3H/^3He$, 1)

The aquifer is composed of glacial sand, silt, clay, and gravel.

Major dissolved-gas data were available for 29 sites. Of these 29 sites, 22 were oxic and 7 were suboxic.

Age interpretations from tracer concentrations were made assuming that recharge elevation was equal to the elevation of the water table. Estimates of recharge temperature and excess air were based on major dissolved-gas data, with recharge temperature and excess air at suboxic sites being constrained using median excess air at oxic sites.

$^3H/^3He$ age was calculated for one site and did not require a correction for terrigenic He.

The raw tracer data, major dissolved-gas data, the ancillary chemical and well construction data that were used in the interpretations, and the piston-flow ages are presented in table B53.

- Advantages associated with these samples:

  - Multiple tracers (SF_6 and $^3H/^3He$, as well as major dissolved gases).

  - Mixture of domestic and monitoring wells, therefore low pumping stress.

  - Relatively short open intervals ranging from 2.66 to 20 (with most <5) feet so mixing likely minimized.

  - Median penetration of center of open interval into water table was 2.44 feet (sampling close to the water table, potentially minimizes mixing).

- Disadvantages associated with these samples:

  - None.

- Depth to water (can affect tracer transport to water table):

  - Median: 16.07 feet

  - Mean: 19.96 feet

  - Min: 4.57 feet

  - Max: 44.28 feet

- Brief analysis:

  The SF_6-based age gradient for these sites is shown in figure B180. The age gradient shows a great deal of scatter, which because of the excellent SF_6-based ^3H reconstruction, would indicate that the scatter may simply result from the location of wells in a flow system, with wells in discharge areas having shallow depths and older ages.

  The reconstructed ^3H plot for SF_6-based ages is shown in figure B181. The reconstruction shows evidence of unmixed, piston-flow transport.

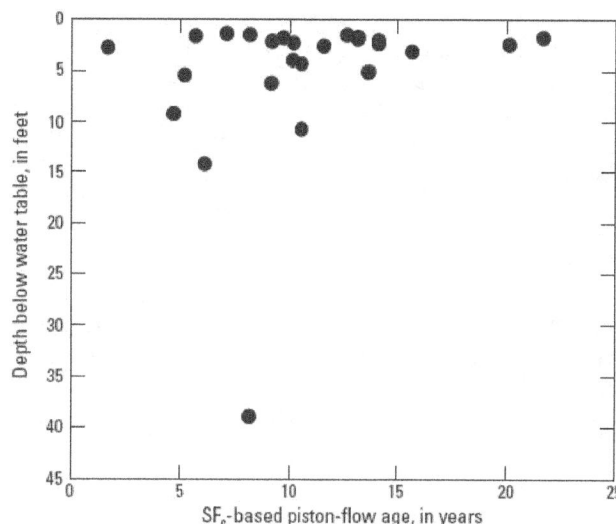

Figure B180. SF_6-based age gradient for dated sites from the LUSCR1 network, UMIS Study Unit.

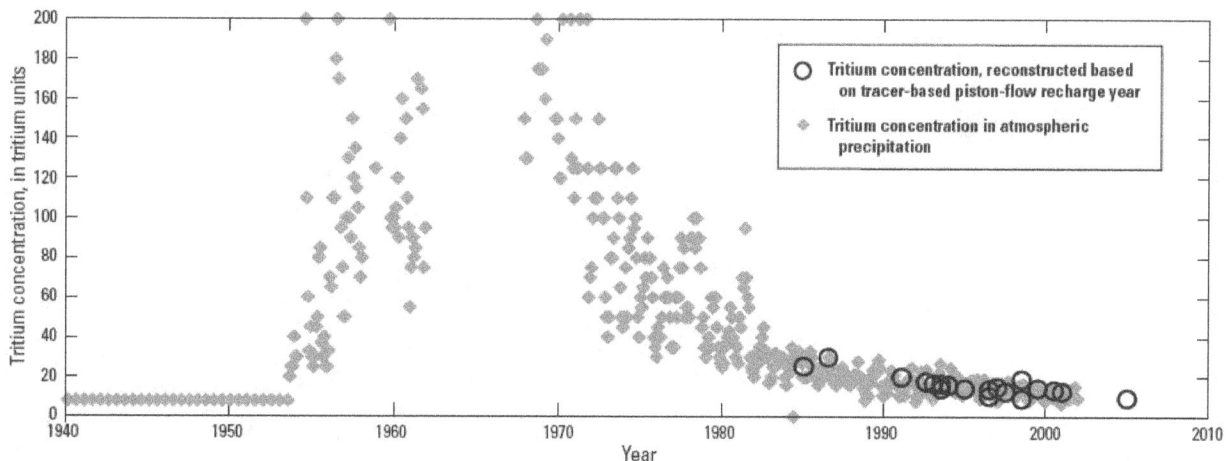

Figure B181.   Reconstructed tritium concentrations (using SF_6-based ages) and tritium in atmospheric precipitation, LUSCR1 network, UMIS Study Unit.

UMIS LUSRC1, REFFO1, and REFFO2

Samples from 27 sites in the UMIS Study Unit were collected in 2006 and 2008 for SF_6 and $^3H/^3He$ (networks and, in parentheses, number of sites):

. LUSRC1 (CFC, 2; SF_6, 21; $^3H/^3He$, 14)

. REFFO1 (SF_6, 1)

. REFFO2 (SF_6, 1)

The aquifer is composed of glacial sand, silt, clay, and gravel.

Major dissolved-gas data were available for 30 sites. Of these 30 sites, 11 were oxic and 19 were suboxic.

Age interpretations from tracer concentrations were made assuming that recharge elevation was equal to the elevation of the water table. Estimates of recharge temperature and excess air were based on major dissolved-gas data, with recharge temperature and excess air at suboxic sites being constrained using median excess air at oxic sites.

$^3H/^3He$ ages were calculated for eight sites (two of the eight sites required a correction for terrigenic helium), while six sites were not datable because of fractionation.

The raw tracer data, major dissolved-gas data, the ancillary chemical and well construction data that were used in the interpretations, and the piston-flow ages are presented in table B54.

. Advantages associated with these samples:

    Multiple tracers (SF_6 and $^3H/^3He$, as well as major dissolved gases).

    Monitoring wells, therefore low pumping stress.

    Relatively short open intervals ranging from 2 to 5 feet so mixing likely minimized.

Median penetration of center of open interval into water table was 3.73 feet (sampling close to the water table, potentially minimizes mixing).

. Disadvantages associated with these samples:

. Suboxic conditions.

. Depth to water (can affect tracer transport to water table):

    Median: 10.04 feet

    Mean: 11.29 feet

    Min: 2.25 feet

    Max: 24.03 feet

. Brief analysis:

The SF_6- and $^3H/^3He$-based age gradients for these sites are shown in figures B182 and B183. The age gradients show a great deal of scatter and may result from SF_6 loss (and therefore older ages) from stripping due to high methane concentrations. In addition, differences in screen length, recharge source/strength, aquifer heterogeneity, pumping stresses, and the position of the well within the flow system may cause some wells to deviate from the general pattern of increasing age with depth.

The reconstructed ^3H plots for SF_6- and $^3H/^3He$-based ages are shown in figures B184 and B185. The reconstructions show evidence of unmixed, piston-flow transport, however, the SF_6 ages are biased old likely as a result of stripping due to high methane concentrations.

The $^3H/^3He$- versus SF_6-based age comparison for this network is shown in figure B186. The age comparison is poor as a result of the SF_6 ages being biased old due to stripping from high methane concentrations.

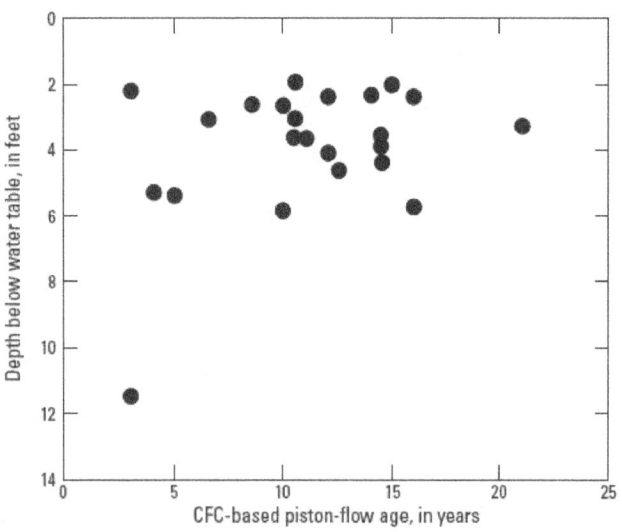

Figure B182. SF_6-based age gradient for dated sites from the LUSRC1, REFFO1, and REFFO2 networks, UMIS Study Unit.

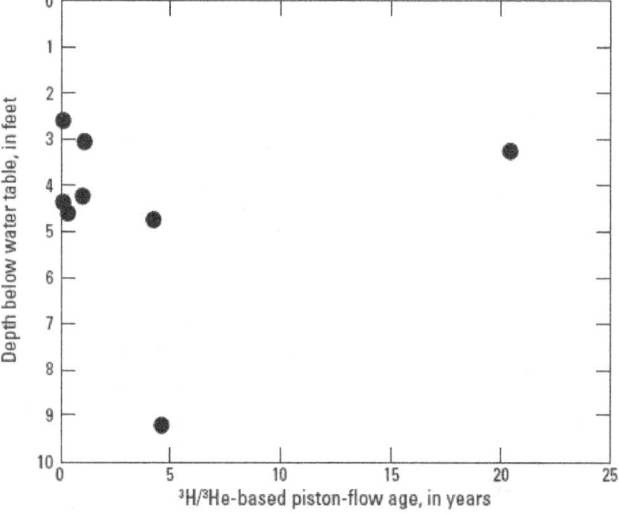

Figure B183. $^3H/^3He$-based age gradient for dated sites from the LUSRC1 and REFFO1 networks, UMIS Study Unit.

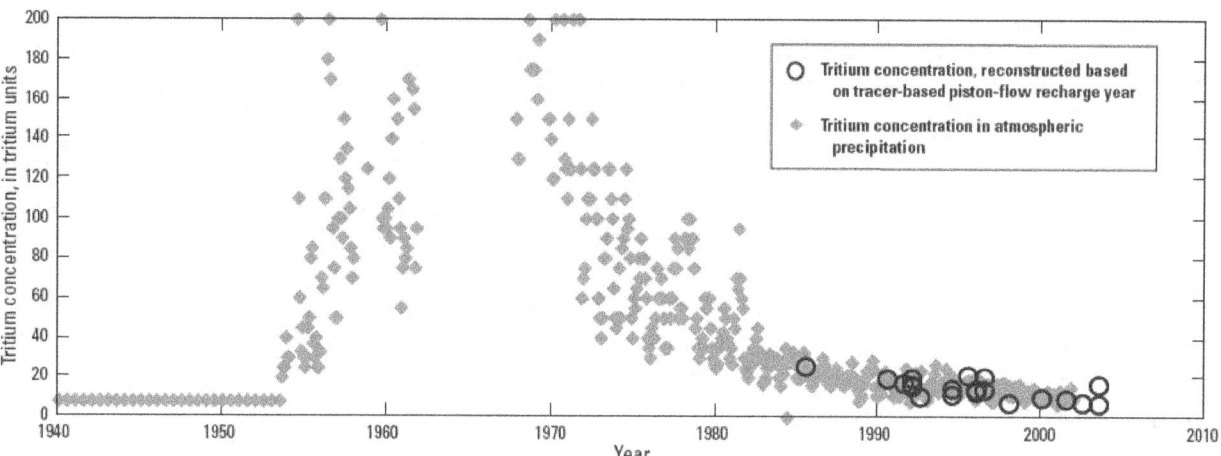

Figure B184. Reconstructed tritium concentrations (using SF₆-based ages) and tritium in atmospheric precipitation, LUSRC1, REFFO1, and REFFO2 networks, UMIS Study Unit.

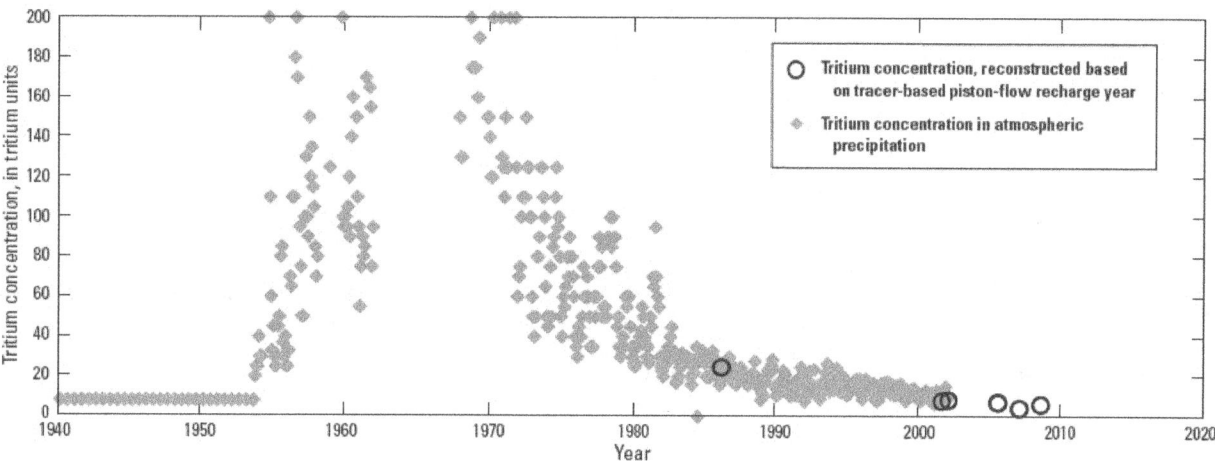

Figure B185. Reconstructed tritium concentrations (using ³H/³He-based ages) and tritium in atmospheric precipitation, LUSRC1 and REFFO1 networks, UMIS Study Unit.

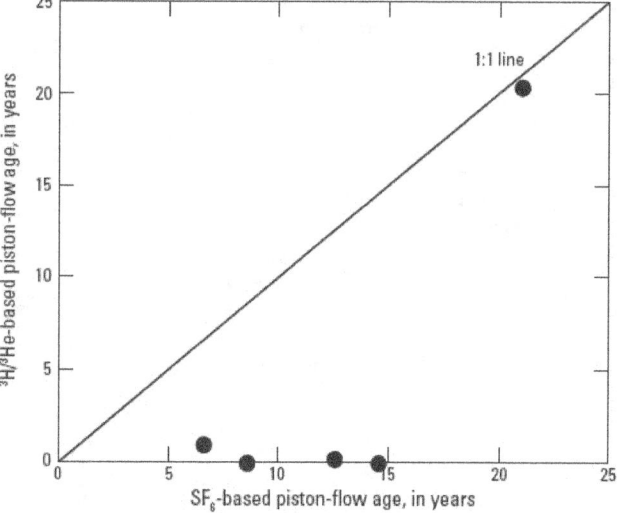

Figure B186. ³H/³He- versus SF₆-based age comparison, LUSRC1 and REFFO1 networks, UMIS Study Unit.

UMIS SUS3

Samples from 20 sites in the UMIS Study Unit were collected in 2007 for SF₆ (networks and, in parentheses, number of sites):

. SUS3 (20)

The sites are finished in aquifers composed of sandstone, limestone, and dolomite.

Major dissolved-gas data were available for 30 sites. Of these 30 sites, 14 were oxic and 16 were suboxic.

Age interpretations from tracer concentrations were made assuming that recharge elevation was equal to the elevation of the water table. Estimates of recharge temperature and excess air were based on major dissolved-gas data, with recharge temperature and excess air at suboxic sites being constrained using median excess air at oxic sites.

The raw tracer data, major dissolved-gas data, the ancillary chemical and well construction data that were used in the interpretations, and the piston-flow ages are presented in table B55.

. Advantages associated with these samples:

. SF_6, as well as major dissolved gases.

. Domestic wells (with some unknown), therefore low pumping stress.

. Disadvantages associated with these samples:

. Relatively large open intervals ranging from 1 to 71.55 feet so mixing likely.

. Median penetration of center of open interval into water table was 69.00 feet (not sampling close to the water table, potentially mixing).

. Depth to water (can affect tracer transport to water table):

. Median: 65.51 feet

. Mean: 76.52 feet

. Min: 11.65 feet

. Max: 219.13 feet

. Brief analysis:

. The SF_6-based age gradient for these sites is shown in figure B187. The age gradient shows no particular structure. Differences in screen length, recharge source/strength, aquifer heterogeneity, pumping stresses, and the position of the well within the flow system may cause some wells to deviate from the general pattern of increasing age with depth.

The reconstructed ^3H plot for SF_6-based ages is shown in figure B188. The reconstruction shows evidence of unmixed, piston-flow transport for some samples, and mixing for other samples, as would be expected for samples taken from wells with large open intervals. Also, several samples plot to the right of the ^3H reconstruction as might be expected for samples taken from sites with relatively deep unsaturated zones.

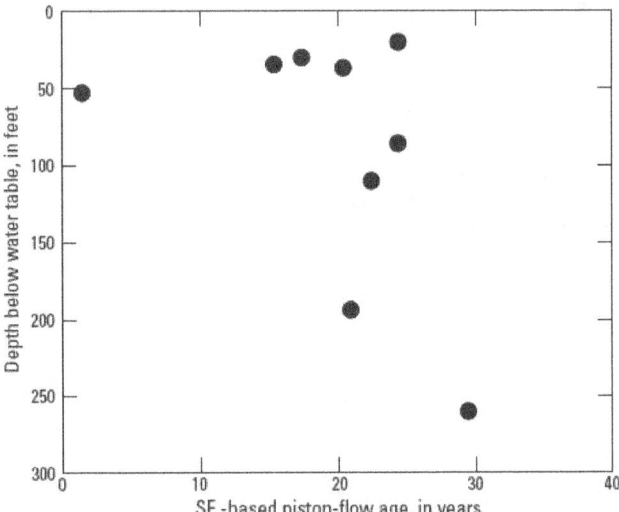

Figure B187. SF_6-based age gradient for dated sites from the SUS3 network, UMIS Study Unit.

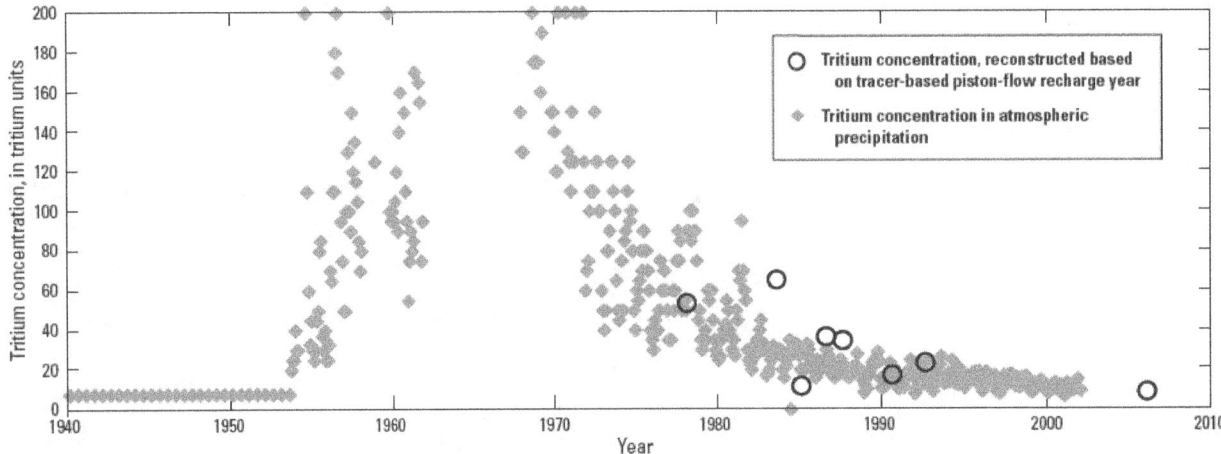

Figure B188. Reconstructed tritium concentrations (using SF_6-based ages) and tritium in atmospheric precipitation, SUS3 network, UMIS Study Unit.

USNK LUSCR2 and LUSCR3

Samples from 14 sites in the USNK Study Unit were collected in 2008 for CFCs, SF_6, and $^3H/^3He$ (networks and, in parentheses, number of sites):

. LUSCR2 (CFCs, 7; SF_6, 7; $^3H/^3He$, 6)

. LUSCR3 (CFCs. 7; and SF_6, 6; $^3H/^3He$, 7)

The aquifer is composed of Snake River Group basalts. Major dissolved-gas data were available for all 14 sites. Of these 14 sites, all 14 were oxic.

Age interpretations from tracer concentrations were made assuming that recharge elevation was equal to the elevation of the water table. Estimates of recharge temperature and excess air were based on major dissolved-gas data.

$^3H/^3He$ ages were calculated for 12 sites (9 of the 12 sites required a correction for terrigenic helium), while the sample from 1 site was lost due to high pressure. No tritium errors were reported so a value of 0.3 TU was used in the $^3H/^3He$ worksheet.

The raw tracer data, major dissolved-gas data, the ancillary chemical and well construction data that were used in the interpretations, and the piston-flow ages are presented in table B56.

. Advantages associated with these samples:

. Multiple tracers (CFCs, SF_6, and $^3H/^3He$, as well as major dissolved gases).

. Domestic wells, therefore low pumping stress.

. Median penetration of center of open interval into water table was 28.25 feet (sampling relatively close to the water table, potentially minimizing mixing).

. Disadvantages associated with these samples:

. Relatively large open intervals ranging from 5.6 to 107.51 feet so mixing likely.

. Depth to water (can affect tracer transport to water table):

. Median: 227.87 feet

. Mean: 241.02 feet

. Min: 113.58 feet

. Max: 453.67 feet

. Brief analysis:

. The CFC-, SF_6-, and $^3H/^3He$-based age gradients for these sites are shown in figures B189, B190, and B191. The age gradients show a great deal of scatter as would be expected for samples taken from domestic wells with large open intervals dispersed among a large geographic area. The CFC- and SF_6-based ages are similar and are confined to a small range in ages, while the $^3H/^3He$-based ages show a wider spread in values. Surprisingly, there was no effect of mantle helium

in these wells, which are finished in basalt, and the $^3H/^3He$-based ages appear to be more reliable than the CFC- and SF_6-based ages as seen in the reconstructed ^3H plots for CFC-, SF_6-, and $^3H/^3He$-based ages, which are shown in figures B192, B193, and B194, and the age comparisons as shown in figures B195, B196, and B197. In addition, the reconstructed ^3H plots show that some wells are a mixture of old and young groundwater.

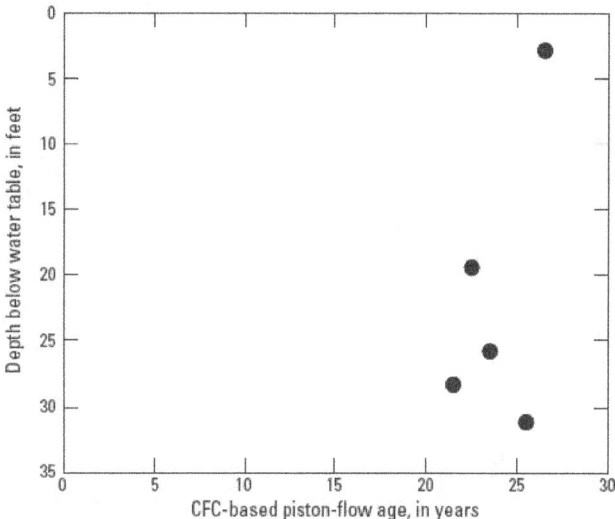

Figure B189. CFC-based age gradient for dated sites from the LUSCR2 and LUSCR3 networks, USNK Study Unit.

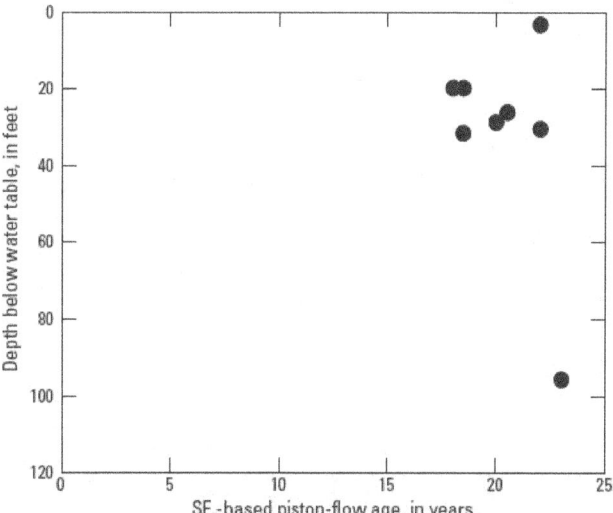

Figure B190. SF_6-based age gradient for dated sites from the LUSCR2 and LUSCR3 networks, USNK Study Unit.

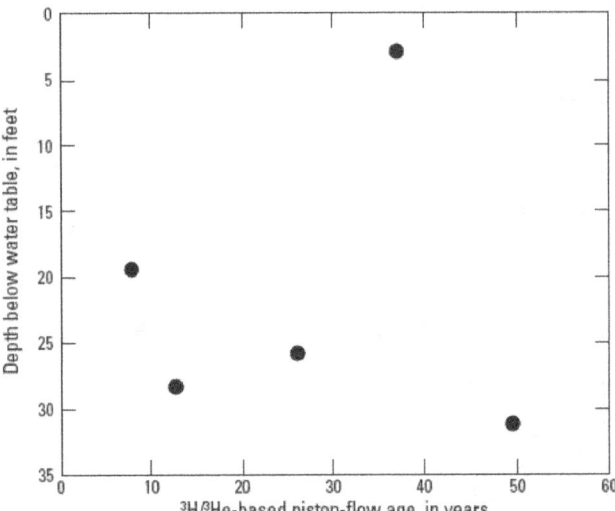

Figure B191. $^3H/^3He$-based age gradient for dated sites from the LUSCR2 and LUSCR3 networks, USNK Study Unit.

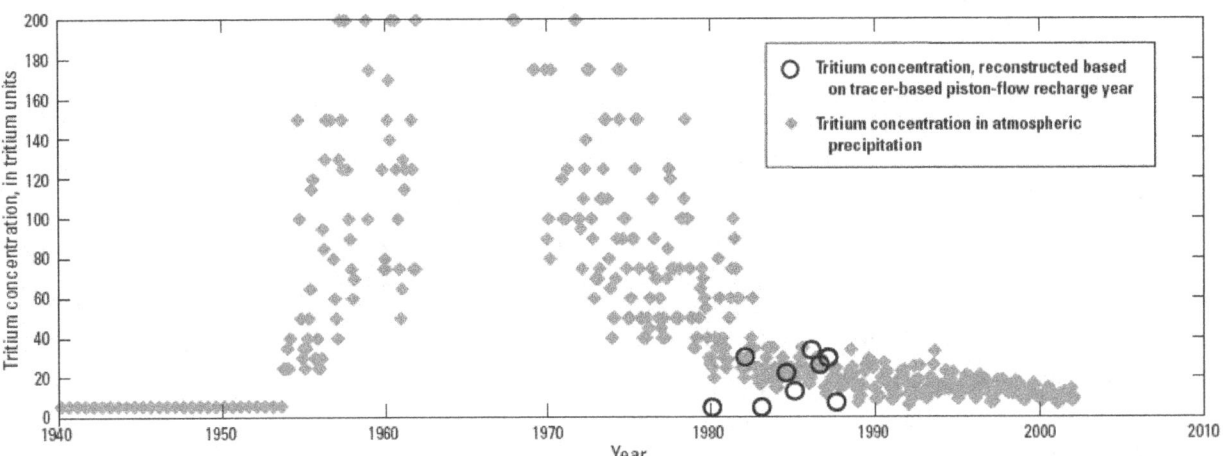

Figure B192.   Reconstructed tritium concentrations (using CFC-based ages) and tritium in atmospheric precipitation, LUSCR2 and LUSCR3 networks, USNK Study Unit.

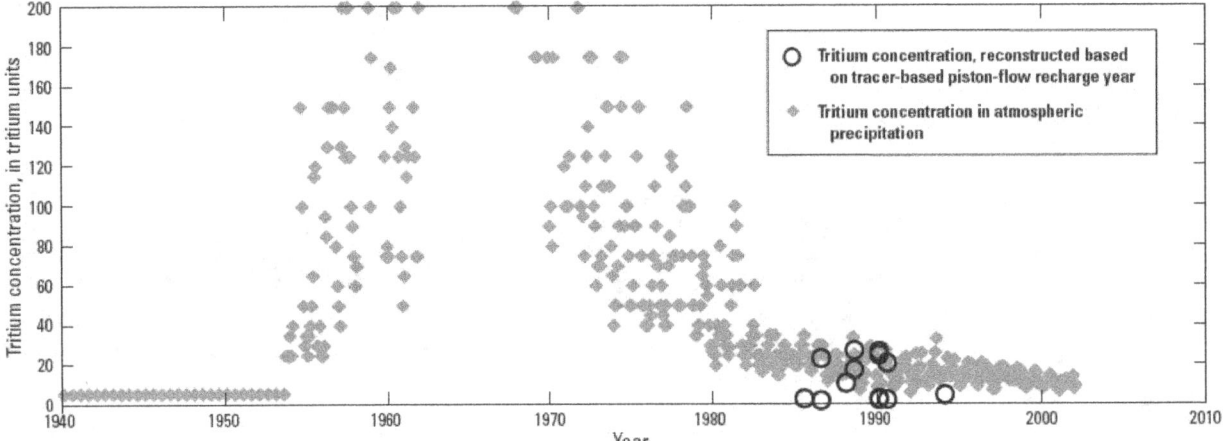

Figure B193.   Reconstructed tritium concentrations (using SF_6-based ages) and tritium in atmospheric precipitation, LUSCR2 and LUSCR3 networks, USNK Study Unit.

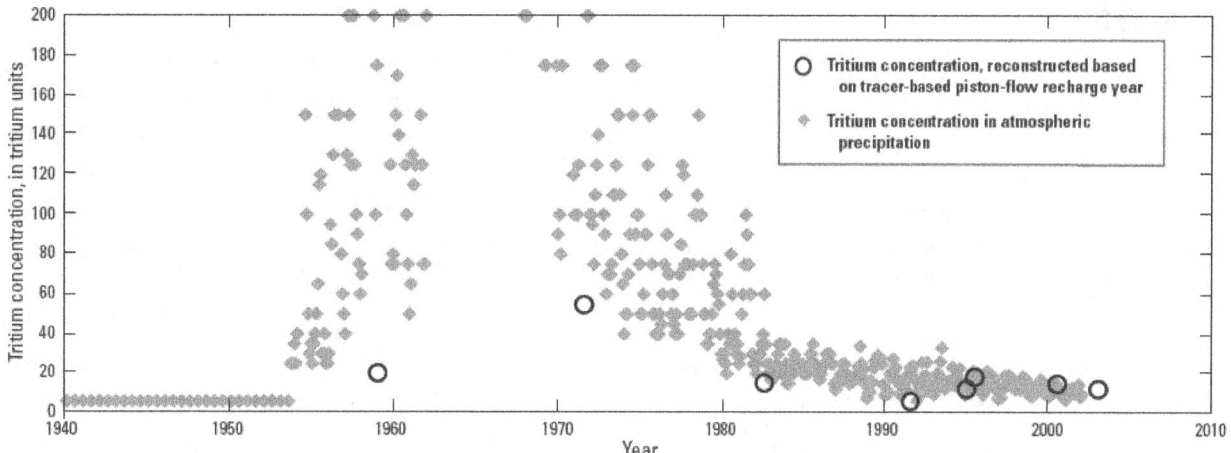

Figure B194.   Reconstructed tritium concentrations (using ³H/³He-based ages) and tritium in atmospheric precipitation, LUSCR2 and LUSCR3 networks, USNK Study Unit.

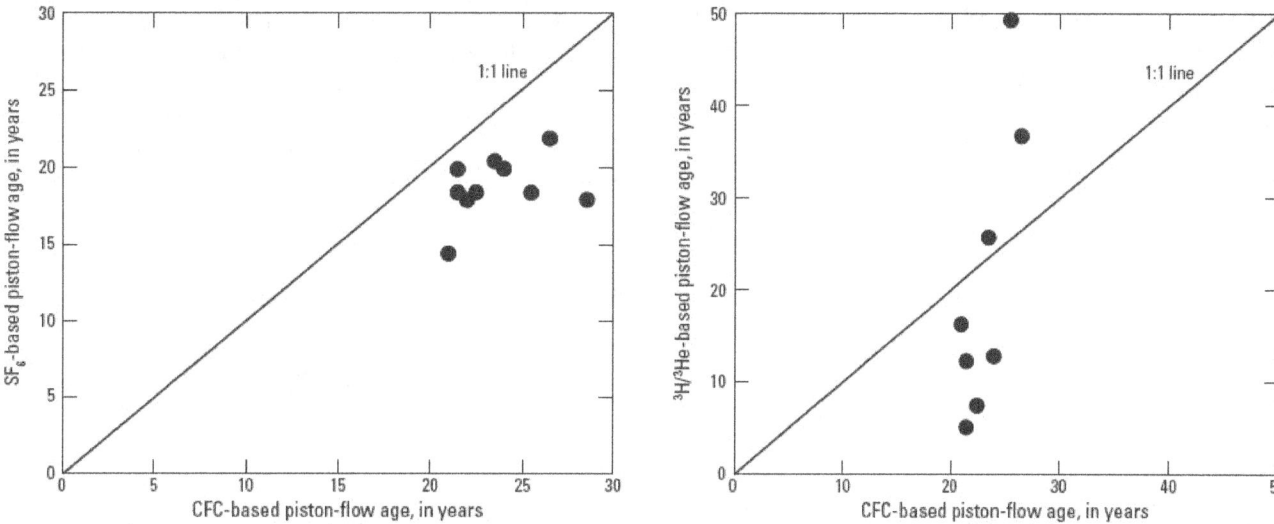

Figure B195.   SF₆- versus CFC-based age comparison, LUSCR2 and LUSCR3 networks, USNK Study Unit.

Figure B196.   ³H/³He- versus CFC-based age comparison, LUSCR2 and LUSCR3 networks, USNK Study Unit.

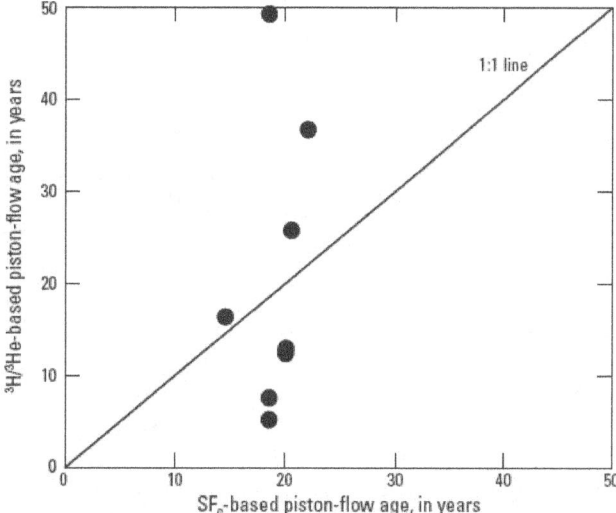

Figure B197.   ³H/³He- versus SF₆-based age comparison, LUSCR2 and LUSCR3 networks, USNK Study Unit.

WMIC REFF01

Samples from seven sites in the WMIC Study Unit were collected in 2007 for CFCs (networks and, in parentheses, number of sites):

. REFFO1 (CFCs, 2)

The aquifer is composed of sand, gravel, and clay.

Major dissolved-gas data were available for seven sites. Of these seven sites, six were oxic and one was suboxic, but recharge temperatures unreasonably high.

Age interpretations from tracer concentrations were made assuming that recharge elevation was equal to the elevation of the water table, that recharge temperature was equal to the mean annual air temperature +1°C, and that excess air concentrations were 2 cc STP/kg.

The raw tracer data, major dissolved-gas data, the ancillary chemical and well construction data that were used in the interpretations, and the piston-flow ages are presented in table B57.

. Advantages associated with these samples:

. CFCs, as well as major dissolved gases.

. Monitoring wells, therefore low pumping stress.

. Relatively short open intervals of 5 feet so mixing likely minimized.

. Median penetration of center of open interval into water table was 14.78 feet (sampling close to the water table, potentially minimizing mixing).

. Depth to water (can affect tracer transport to water table):

. Median: 18.78 feet

. Mean: 25.75 feet

. Min: 12.51 feet

. Max: 50.72 feet

. Brief analysis:

. Only two sites were sampled for CFCs in this network, and there were no tritium data, therefore no plots were created. The deeper sample is older than the shallower sample, but with only two points, it is difficult to determine the age structure of the aquifer.

.